ON GROUP-THEORETIC DECISION PROBLEMS AND THEIR CLASSIFICATION

BY

CHARLES F. MILLER, III

ANNALS OF MATHEMATICS STUDIES
PRINCETON UNIVERSITY PRESS

Annals of Mathematics Studies

Number 68

ON GROUP-THEORETIC DECISION PROBLEMS AND THEIR CLASSIFICATION

BY

CHARLES F. MILLER, III

PRINCETON UNIVERSITY PRESS

AND

UNIVERSITY OF TOKYO PRESS

PRINCETON, NEW JERSEY

1971

LC Card: 70-146647
ISBN: 0-691-0-08091-7
AMS, 1968: 0275, 2010

Published in Japan exclusively by
University of Tokyo Press;
in other parts of the world by
Princeton University Press

Printed in the United States of America

PREFACE

This study is concerned with group-theoretic decision problems – specifically the word problem, conjugacy problem, and isomorphism problem for finitely presented groups. The principal aims of the work are twofold: (1) to prove unsolvability results for decision problems in classes of finitely presented groups which are in some sense "elementary" (for example, residually finite); and (2) to obtain some new unsolvability results for the isomorphism problem. Of course, there is some overlap between these two goals.

The main objective is to present these new results, but a variety of background material has been included, and proofs of several background results are supplied. Although not intended as a textbook, it is hoped that inclusion of background will help make the monograph accessible to a larger mathematical audience. A proof of the unsolvability of the word problem has not been included, but this is now available in at least two textbooks ([44], [48]).

The Introduction (Chapter I) contains formulations of the decision problems considered, a brief survey of the field, and statements of the principal new results. In Chapter II, several useful group-theoretic results concerning equality and conjugacy of words in certain groups are considered. Chapter III is concerned with unsolvability in certain "elementary" groups. In Chapter IV, the more difficult arbitrary r.e. degree results for "elementary" groups are proved, and then applied to the isomorphism problem. Finally, Chapter V is devoted to obtaining a strong form of the unsolvability of the isomorphism problem for finitely presented groups in general.

This work had its origin in my doctoral dissertation [53], written at the University of Illinois.

I wish to thank William W. Boone for his invaluable advice, guidance, and encouragement during the preparation of this work. I would also like to thank K. I. Appel, D. J. Collins, and P. E. Schupp for their interest, encouragement, and suggestions, and to thank J. Mary Tyrer for her help with the proofs. Partial support from the U. S. National Science Foundation is gratefully acknowledged.

Charles F. Miller, III

CONTENTS

On Group-Theoretic Decision Problems and Their Classification

CHAPTER I

INTRODUCTION

The subject of this work is "group-theoretic decision problems" – an area of mathematics lying in the intersection of combinatorial group theory and mathematical logic. The main topics considered are the word problem, the conjugacy problem and the isomorphism problem for groups. All of these problems arose from topological considerations and were originally formulated by M. Dehn [20]. We begin with algebraic formulations of these problems, indicate their topological connections, and state our major results. For group theoretic background we cite the excellent book [30] by Magnus, Karrass, and Solitar; see [43] for logical concepts.

By a presentation of a group we mean an ordered pair $<S; D>$ where S is a set and D a collection of words on the elements of S and their inverses. By the group G presented by $<S; D>$ we mean the quotient group of the free group on S by the normal closure of the words in D. (For further details see [30]). We usually write $G = <S; D>$ in this situation. Except when necessary, we will not distinguish between a group (as an abstract algebraic object) and its presentation (as a notation in some logical system). [1]

$G = <S; D>$ is said to be *finitely generated* (f.g.) if S is finite and *finitely presented* (f.p.) when S and D are both finite. Note that being generated by a finite set of elements or having a finite presentation is an algebraic property of groups (preserved under isomorphism). Finitely pre-

1 There are other possible viewpoints towards presentations. In particular, presentations can be viewed as logical systems – see Boone [9] for this approach.

sented groups arise naturally in topology: they are exactly the fundamental groups of finite simplicial complexes or alternatively the fundamental groups of closed differentiable n-manifolds ($n \geq 4$) (see [14], [42], [46], and [49] for topological connections).

Viewing the group $G = \langle S; D \rangle$ as the quotient of the free group F on S by the normal closure of D, recall that a word $W = 1$ (the empty word) in G if and only if W is equal in F to a product of conjugates of words in D. Moreover, $W_1 = W_2$ in G if and only if there is a word $Z = 1$ in G such that $W_1 = Z W_2$ in F.

In case G is finitely generated, the *word problem* (WP) (or in some languages the identity problem) for G is the algorithmic problem of deciding for arbitrary words W of G whether or not $W = 1$ in G. In case $G = \langle S; D \rangle$ is such that S is finite and D is a recursively enumerable (r.e.) set of words, G is said to be *recursively presented.* [2] In particular, f.p. groups are recursively presented. Observe that the word problem is r.e. for recursively presented groups.

The algorithmic problem of deciding for an arbitrary pair of words $U, V \in G$ whether or not U and V are conjugate in G (i.e. whether or not there exists $W \in G$ such that $U = W^{-1}VW$ in G) is called the *conjugacy problem* (CP) or *transformation problem* for G. For recursively presented groups, the conjugacy problem is r.e. Observe that the word problem for G is one-one reducible to the conjugacy problem for G since $U = 1$ in G if and only if U is conjugate to 1 in G.

Let $\Phi = \{ \Pi_i, i \geq 0 \}$ be a recursive class of finite presentations of groups (say on some fixed alphabet). The *isomorphism problem* (IsoP) for Φ is the algorithmic problem of deciding whether or not $\Pi_i \cong \Pi_j$ for arbitrary i and j (i.e. whether or not Π_i and Π_j are presentations for the same abstract group). The class Φ is required to be recursive so that the algorithmic problem is well-posed.

---

2 By a device due to Craig, a group is recursively presented (in our sense) if and only if it has some presentation $\langle S'; D' \rangle$ where S' is finite and D' is a recursive set of words. See Boone [9] or [10] in this connection.

In the context of topology, the word problem corresponds to the problem of deciding whether a closed path is contractible to a point. The conjugacy problem corresponds to the problem of deciding whether two closed paths are free homotopic. Since homotopy equivalent spaces have isomorphic fundamental groups, the isomorphism problem is related to the problem of homotopy equivalence.

Finally, consider a f.g. group $G = \langle S; D \rangle$ and a set of words $W_1, W_2, \ldots$ of G. Let H be the subgroup of G generated by $W_1, W_2, \ldots$. The *generalized word problem* (GWP) or *membership problem (*or *Magnus problem)* for H in G is the algorithmic problem of deciding whether or not an arbitrary word $U \in G$ belongs to the subgroup H (i.e. whether or not there is a product of the W_i and their inverses which is equal to U in G). If the set $\{W_i\}$ is r.e., then the generalized word problem for H in G is r.e. provided G is recursively presented. Clearly, the word problem for G is just the GWP for the trivial subgroup in G. The GWP as formulated here is often called the *extended word problem,* but we follow [30] in our choice of terminology.

Each of Dehn's three fundamental problems is known to be unsolvable for f.p. groups. Novikov [39] proved the conjugacy problem unsolvable for f.p. groups. Shortly thereafter, Boone [8] and Novikov [40] proved the word problem for f.p. groups is unsolvable (also see [9], [16], and [25] in this connection). Using these results, Adjan [1] and Rabin [41] showed that the isomorphism problem for f.p. groups is unsolvable. Indeed, most significant algorithmic problems concerning all f.p. groups have been shown to be unsolvable (see [5] and [41]). As noted above, the word problem is always reducible to the conjugacy problem. However, Fridman [22] gives an example of a f.p. group with solvable word problem, but unsolvable conjugacy problem.

In view of the Friedberg-Mucnik Theorem (see [43]), it is natural to ask which r.e. Turing degrees contain word problems, etc. The answer is that all r.e. degrees contain such problems. Let D be any given r.e. degree. Bokut [7], Boone [11], Clapham [17], and Fridman [23] have all given constructions which yield a f.p. group with word problem of degree D. Collins [18] gives a construction yielding a f.p. group with solvable word problem, but conjugacy problem of degree D. Finally, Boone [12] has constructed a recursive class of finite presentations with isomorphism problem of degree D. Indeed, it seems that whenever one has an unsolvability result about f.p. groups, there should also be an arbitrary r.e. degree analogue (usually with a much harder proof).

Let G be a finitely presented group and H a finitely generated subgroup of G generated by words $U_1, \ldots, U_n$. The set of all words in the U_i is r.e., and the set of all words G which are equal to 1 in G is r.e. Hence the intersection of these two sets is r.e. – that is the set of words in the U_i which are equal to 1 in G is r.e. Thus H is a recursively presented group. A remarkable result of Graham Higman [25] (see also [48]) asserts the converse holds. *Higman's Theorem* is the following: A finitely generated group H can be embedded in a finitely presented group if and only if H is recursively presented. The existence of a f.p. group with unsolvable word problem follows easily from Higman's Theorem.

So far we have mentioned mostly unsolvability results. There are, however, many positive results concerning word and conjugacy problems. Before stating some of these, we recall some definitions and facts from group theory.

A property P of groups is called a *poly-property* if, whenever N and G/N have the property P, so has G. Here N is a normal subgroup of G.

LEMMA 1: *The following are poly-properties:*

(1) being finitely generated

(2) having a finite presentation

(3) satisfying the maximum condition for subgroups

(4) being finitely generated and having a solvable word problem.

That (1)-(3) are poly-properties is shown in P. Hall [51], and that (4) is a poly-property is easily verified.

A group G is *poly-P* if it can be obtained from the trivial group by a finite succession of extensions by groups in P, i.e. if there is a finite series of subgroups

$$G = G_0 \supset G_1 \supset \ldots \supset G_n = \{1\}$$

such that G_{i+1} is normal in G_i and $G_i/G_{i+1} \in P$. As an application of Lemma 1, it follows that polycyclic groups are finitely presented, satisfy the maximum condition for subgroups, and have solvable word problem. Observe that solvable groups are just the polyabelian groups.

Let P be a property of groups. A group G is said to be *residually* P if for every $1 \neq W \in G$ there is a normal subgroup N_W of G such that $W \notin N_W$ and G/N_W has the property P. Equivalently, G is residually P if the intersection of the normal subgroups N of G such that G/N has P is the identity. For instance, free groups are residually nilpotent and residually finite (see [30]). Baumslag [3] has shown that the automorphism group of a f.g. residually finite group is again residually finite.

A group G is called *hopfian* if $N \lhd G$ and $G/N \cong G$ imply that $N = \{1\}$. Equivalently, G is hopfian if every epic endomorphism of G is an automorphism. F.g. residually finite groups are hopfian (see [38] 41.44).

Dyson [21] and Mostowski [35] have shown that f.p. residually finite groups have solvable WP. Briefly, the argument is as follows: Let G be f.p. residually finite, and let W be a word of G. Then $W \neq 1$ in G if and only if $W \neq 1$ in some finite quotient of G. Since G is f.p., one can construct a list of all finite quotients of G and check to see whether or not $W \neq 1$ in successive quotients. Hence, one can enumerate $\{W | W \neq 1 \text{ in } G\}$. On the other hand, since G is f.p., $\{W | W = 1 \text{ in } G\}$ can be enumerated. Since every word of G appears in exactly one of these lists, this solves the word problem for G.

Baumslag [4] has asked whether f.p. residually finite groups have solvable conjugacy problem – we answer this question in the negative. Moreover, we show the isomorphism problem for f.p. residually finite groups is unsolvable.

Any f.g. residually finite group is hopfian. Schupp and the author [34] have shown that any f.p. group G can be embedded in a f.p. hopfian group H in such a way that the WP for G is (Turing) equivalent to the WP for H. It follows that the WP for f.p. hopfian groups can be unsolvable.

F.g. nilpotent groups are polycyclic and so are f.p. Hirsch has shown polycyclic groups are residually finite (see [38] 32.1). In particular, f.g. nilpotent groups and polycyclic groups have solvable word problems (this already followed from Lemma 1). Blackburn [6] has shown the conjugacy problem for f.g. nilpotent groups is solvable. Recently, Remeslennikov [50] has solved the conjugacy problem for polycyclic groups. Indeed he has shown, as Blackburn did for f.g. nilpotent groups, that polycyclic groups are *conjugacy separable*. (G is conjugacy separable means that if U, V are not conjugate in G then their images in some finite quotient of G are not conjugate. If G is conjugacy separable and f.p., then G has solvable

Toh [57] has shown that polycyclic groups are *subgroup separable,* i.e. if H is a subgroup of the polycyclic group G and $W \notin H$, then there exists an epimorphism ϕ: $G \to F$ where F is a finite group such that $\phi(W) \notin \phi(H)$. Thus, by an argument similar to that showing the WP is solvable for f.p. residually finite groups, Toh has shown the GWP is solvable for polycyclic groups.

That the WP, CP, GWP, IsoP are solvable for f.g. abelian groups follows easily from the "fundamental theorem for f.g. abelian groups" (see for instance [30] p. 146). Clearly, f.g. abelian groups are f.p. and nilpotent.

Many other positive results are known, and we have mentioned only those which have some bearing on this work. The survey articles [13], [54], and [56] contain additional information on group-theoretic decision problems.

Aims of this work:

The word problem, conjugacy problem, and other decision problems are known to be unsolvable for finitely presented groups in general. Our aims are twofold:

First, to prove unsolvability results for these decision problems in classes of finitely presented groups which are in some sense "elementary." Two methods of "measuring" the elementary nature of finitely presented groups will be considered: (1) algebraic constructions – how a group is built from free groups by algebraic constructions such as free products with amalgamation; and (2) algebraic classes – groups defined by algebraic properties such as residual finiteness and residual freeness.

Second, to prove some new unsolvability results for the isomorphism problem for finitely presented groups. Of course, there is some overlap between these two goals, and the isomorphism problem for certain elementary classes is shown to be unsolvable.

As we have indicated, a major goal of this work is to find unsolvable problems in certain "elementary" groups. Most of our constructions exploit a well-known result of Higman, Neumann, and Neumann [26] (see also [27] pp. 53-54) on forming extensions by adding inner automorphisms. We have chosen to call these "Britton extensions" for two reasons: (1) fundamental use is made of Britton's Lemma which characterizes words equal to 1 in such extensions; and (2) it seems easier to say "Britton extension" than "Higman-Neumann-Neumann extension." In any event, those authors are due an apology for our choice of terminology.

In "measuring" the elementary nature of a group by how it is built from free groups, the following algebraic constructions will be considered:

(1) free products with amalgamated subgroup

(2) Britton extensions

(3) split extensions

(4) direct products.

From the viewpoint of decision problems, free groups are rather pleasant. Namely, the word problem, the conjugacy problem, and the generalized word problem for finitely generated subgroups are all recursively solvable (see [30]).

Each of the above constructions occurs quite frequently in infinite group theory. If a f.p. group is built from free groups by one application of these constructions, it is in a sense "elementary."

Algebraic constructions provide a link between the word problem and the generalized word problem. For suppose H is a f.p. group which has a f.g. subgroup A with unsolvable GWP (note that H may have solvable WP). Let H_1 be another copy of H with corresponding subgroup A_1. Now form the free product with amalgamation $G = (H * H_1;\ U = U_1 \text{ for all } U \in A)$. Now if W is an arbitrary word of H and W_1 the corresponding word in H_1, then $W W_1^{-1} = 1$ in G if and only $W \in A$. Since the GWP for A in H is unsolvable, this shows the WP for the f.p. group G is unsolvable (even though H may have had solvable WP). Similar considerations apply to Britton extensions.

In Chapter II, our proof [32] of Britton's Theorem A is given. The intimate relationships between Britton extensions and free products with amalgamation are investigated. A Theorem of Solitar is quoted, and a generalization of a lemma due to Collins is proved. Complete understanding of Chapter II is assumed in the remaining chapters. The proofs of our main results occupy Chapters III thru V.

Most of our constructions presuppose the existence of a f.p. group H with word problem of arbitrary r.e. degree D. The existence of such groups is, of course, one of the major results in the field of group-theoretic decision problems ([7], [11], [17], [23]). The reader need *not* be familiar with a proof of this result. Our "plot" is to take such a group H and construct from it elementary groups with unsolvable problems of various kinds.

Statements of Results:

The conjugacy and word problems for elementary groups:

Several unsolvability results for conjugacy and word problems in particularly "elementary" groups are obtained. By suitably modifying an example of Boone [10], the following is established: (Theorem IV - 1)

(i) Let $D_1 \leq D_2$ be given r.e. degrees of unsolvability. Then there is a finitely generated, recursively presented group G with word problem of degree D_1 and conjugacy problem of degree D_2. In fact, G is the free product with (infinitely generated) amalgamation of two finitely generated free groups.

Next the presence of unsolvable problems in "elementary" finitely presented groups is considered. A finitely presented Britton extension of a free group or the free product of two free groups with finitely generated amalgamation must have a solvable word problem. However, even at this elementary level the conjugacy problem can be unsolvable: Let H be a given finitely presented group with word problem of degree D. A uniform method is given for constructing from H the following: (Theorem IV - 12)
(ii) A finitely presented Britton extension G of a free group F which has solvable word problem, but conjugacy problem of degree D.
(iii) A group K which is the free product with finitely generated amalgamation of two free groups and which has solvable word problem, but conjugacy problem of degree D.

In [18] Collins constructed a f.p. group with solvable word problem, but conjugacy problem of degree D. However, his groups are not as "elementary" as those in our construction – a mild generalization of his methods are used in our proofs. Finally, it is shown (Theorem V - 12) that there is a two stage, finitely presented Britton extension of a free group with word problem of degree D. This is the most elementary stage at which an unsolvable word problem could occur.

The method of transferring the word problem for H into the conjugacy problem for G of (ii) in a degree-preserving way provides a link in the program of constructing inter-reducibilities among algebraic decision problems as initiated by Singletary.

Unsolvable problems in the automorphism group of a free group:

Let D be a given r.e. degree of unsolvability, and let F be a free group of finite rank ≥ 3. Let A be the automorphism group of F. The group A is residually finite and has a solvable word problem. Several finite presentations are known for A. By exploiting techniques used in proving (ii) above, we prove the following two results:
(iv) There is a finitely generated subgroup B of A whose membership problem (generalized word problem or Magnus problem) has degree D.

Moreover, if D is suitably given, a finite set of generators for B can be effectively found. (Theorem III-11).

(v) If rank F $\geq$ 4, there is a finitely generated subgroup C of A which has unsolvable conjugacy problem of degree $\geq$ D (Theorem III-16).

Unfortunately, the subgroup C in (v) is not finitely related; however, since A is residually finite, it follows that C is residually finite. Hence there is a finitely generated, recursively presented, residually finite group with unsolvable conjugacy problem.

A finitely presented, residually finite group with unsolvable conjugacy problem:

One of the finitely presented groups with unsolvable conjugacy problem constructed above is found to be a split extension of one finitely generated free group by another. In general, one can show (Theorem III-7):

(vi) Suppose that $1 \to A \to E \to B \to 1$ is an exact sequence of groups where A and B are residually finite and A is f.g. If either A has trivial center or the sequence splits, then E is residually finite. Since free groups are residually finite this proves (Theorem III-9):

(vii) There is a finitely presented, residually finite group with unsolvable conjugacy problem. This answers a question of Baumslag [4]. It is known that finitely presented, residually finite groups have solvable word problem. Moreover, using (vi) and (ii) one can show (Theorem IV-12):

(viii) For each r.e. degree D, there is a finitely presented, residually finite group with conjugacy problem of degree D.

In section IV-C, several applications of this result are given. For example, it is shown that having a solvable conjugacy problem is not hereditary in the class of f.p., residually finite groups. This answers a question of Collins [19].

Unsolvable problems in direct products of free groups:

Let F be a free group of finite rank ≥ 2. The direct product $F \times F$ is then a f.p. residually free group. Hence $F \times F$ is also residually nilpotent and residually finite. In [52] Mihailova has shown the generalized word problem for $F \times F$ is unsolvable (see Theorem III-17). By using her construction, we obtain the following additional results:

(ix) $F \times F$ has a finitely generated subgroup L such that: (1) L has unsolvable conjugacy problem and (2) the generalized word problem for L in G is unsolvable (Theorem III-23). Now L is also residually free, so a recursively presented residually free group can have unsolvable conjugacy problem.

(x) For F of rank at least 9, the problem to determine of an arbitrary finite set of words whether or not they generate $F \times F$ is recursively unsolvable.

Since $F \times F$ can be "nicely" embedded in suitable unimodular groups, these results have analogues for unimodular groups (see section III-C).

On the isomorphism problem for groups:

The results obtained above for elementary groups can be applied to prove the following: (Corollary III-27)

(xi) There is a recursive class $\Omega = \{\Pi_i, i \geq 0\}$ of finite presentations of groups such that (1) each Π_i is residually finite; (2) each Π_i is a free product with finitely generated amalgamation of two free groups; and (3) the isomorphism problem for Ω is recursively unsolvable.

The following result improves upon results of Rabin [41] and Boone [12] on the isomorphism problem for groups: (Theorem V-2):

(xii) Let $E(i,j)$ be an equivalence relation on the natural numbers N. Then E is r.e. $\leftrightarrow$ there is a recursive class $\Phi = \{\Pi_i, i \in N\}$ of finite presentations of groups such that $\Pi_i \cong \Pi_j$ if and only if $E(i,j)$. In particular, taking E to be equality of words in a finitely presented group G, there is a recursive class $\Phi = \{\Pi_u, u \in G\}$ of finite presentations of groups such that $\Pi_u \cong \Pi_v$ if and only if $u = v$ in G.

This result has the following corollary:

(xiii) Every r.e. many-one degree of unsolvability contains an isomorphism problem, but the one-one degree of a simple set does not (Corollary V-3).

On a problem of Graham Higman:

Higman [25] has shown the existence of a universal finitely presented group – one which simultaneously embeds all finitely presented groups.

However, using results of Boone and Rogers [15] one can show the followin (Section III-D):

(xiv) There does *not* exist a finitely presented group with solvable word problem which contains an isomorphic copy of every finitely presented group with solvable word problem. This answers a question of Graham Higman.

Notational Matters: The following notational conventions will be used:

$< s_1, \ldots, s_n; R_1, \ldots, R_m >$ denotes the group presented by generators $s_1, \ldots, s_m$, subject to the defining relations $R_1 = 1, \ldots, R_m = 1$. Moreover, $< s_1, \ldots, s_n >$ will denote the free group on $s_1, \ldots, s_n$ and, if Q is a set of words in some group, $< Q >$ will denote the subgroup generated by Q. Generally, lower case letters will be used for generators of a group, while uppercase letters will be used for groups or for words in some specified group. $G \cong H$ means the groups G and H are isomorphic. If U and V are words of G, then $U \equiv V$ means U and V are identical as words, $U \underset{G}{=} V$ or simply $U = V$ means that U and V are equal in the group G, and $U \underset{G}{\sim} V$ or simply $U \sim V$ means that U and V are conjugate in G. The empty word is denoted by 1 as is the trivial group. The symbol $\approx$ will denote an equivalence relation (see Chapter II).

Let G and H be groups. $G \times H$ denotes the direct product of G and H, while $G * H$ denotes their (ordinary) free product. $(G * H; A = B, \phi)$ denotes the free product of G and H amalgamating their respective subgroups A and B via the isomorphism ϕ. When ϕ is specified or obvious from the context we simply write $(G * H; A = B)$.

Theorems and lemmas are numbered consecutively within each chapter; thus Lemma 5 would be the item following Theorem 4. In a different chapter, Lemma 4 of Chapter II will be cited as Lemma II-4. The following results are unnumbered and bear names from the literature: Theorem A, Britton's Lemma, Collins' Lemma, Solitar's Theorem, Rabin's Theorem, and Higman's Theorem.

CHAPTER II

PROPERTIES OF BRITTON EXTENSIONS

A. BRITTON'S THEOREM A

In [1] Britton gives a "group-theoretic" proof of the unsolvability of the word problem for finitely presented groups. His proof uses a very powerful result [16, Lemma 4 (the principal lemma)] on presentations which has become known as *Britton's Lemma* and has since been used by several other authors. In the appendix to [16] Britton states a generalization of this lemma as *Theorem A*. Britton briefly indicates how Theorem A can be deduced from the previous lemma and how it may be further generalized. We give a direct proof of Theorem A in its most general form which specializes immediately to Britton's Lemma. Our proof appeals more directly to the theory of free products with amalgamation.

The following lemma is an immediate consequence of the normal form for free products with amalgamation:

LEMMA 1: *Suppose* $G = (A_1 * A_2; H_1 = H_2, \phi)$ *and let*

$$W \equiv XY_1 \dots Y_n \ (n \geq 1)$$

where $X \in H_1 = H_2$ *and* $Y_i \in A_{v(i)}$, *where* $v(i) \neq v(i+1)$ *so that the* Y_i *lie in alternating factors. Then if* $W = 1$ *in* G *there is some* i *such that* $Y_i \in H_1 = H_2$.

P will denote a set of letters indexed by a set V. We write $p(v)$ rather than p_v to avoid several levels of subscripting. Let $E = < S;D >$ be any presentation and assume P and S are disjoint. A presentation E^*

is said to have *stable letters* P and corresponding *basis* E if it has the form

$$E^* = <S,p;\ D,\ p(y_i)^{-1}\ A_i p(z_i) = B_i (i \in I)>$$

where y_i, $z_i \in V$ and A_i and B_i are words over S.[1]

We write $p(y) \approx p(z)$ if $p(y) = p(z)$ in the free group obtained from E^* by setting all the letters of S equal to 1. Let $K(v) = \{i \in I\colon p(y_i) \approx p(v)\}$ denote the induced equivalence classes in I. A(v) denotes the subgroup of E generated by the A_i such that $i \in K(v)$, and similarly for B(v).

E^* as above satisfies the *generalized isomorphism condition* (GIC) if for each K(v) the map $A_i \to B_i$ $(i \in K(v))$ defines an isomorphism between A(v) and B(v).

Let T,U be words over S and let $y,z \in V$. We say Tp(y) *produces* p(z)U if the word Tp(y) can be transformed into p(z)U by a sequence of operations of the form

$$XA_i p(z_i) Y \to X p(y_i) B_i Y$$

or

$$XA_i^{-1} p(y_i) Y \to X p(z_i) B_i^{-1} Y$$

where X,Y are words over S.

It can be shown that if Tp(y) produces p(z)U then T is a word in A(y), say $T \equiv w(A_i)$ and U the corresponding word $w(B_i)$ in B(y). Moreover $Tp(y) = p(z)U$ in E^*, and $p(y) \approx p(z)$.

A word W *involves* the letter p if either p or p^{-1} is a subword of W. (In particular, the word pp^{-1} involves p).

THEOREM A: *Suppose* E^* *as above satisfies the* GIC. *Then*

(I) E *is embedded in* E^* *and*

(II) *if* $W = 1$ *in* E^* *where* W *involves at least one letter* $p \in P$, *then* W *contains a subword* (1) $p(y)^{-1} C p(z)$ *or* (2) $p(y)\ C\ p(z)^{-1}$ *where* C *is a word over* S. *In case* (1), C *is equal in* E *to a word*

---

1 The concept of stable letters is motivated by a similarity to inner automorphisms and by an analogy to internal states of a Turing machine. See [44] Chapter 12 in this connection.

$w(A_i) \in A(y)$ *and* $w(A_i)\,p(z)$ *produces* $p(y)\,w(B_i)$. *In case* (2), C *is equal in* E *to a word* $w(B_i) \in B(y)$ *and* $w(A_i)\,p(z)$ *produces* $p(y)\,w(B_i)$.

Proof: Let $< a(v)\,(v \in V) >$ denote the free group generated by the a(v). Generally, $< Q >$ denotes the subgroup generated by the set Q. Define the (ordinary) free products:

$$F = < a(v)\,(v \in V) > * E$$
$$G = < b(v)\,(v \in V) > * E$$

Consider the subgroups $M_v = < a(y_i)^{-1} A_i a(z_i)\,(i \in K(v)) >$ of F and $N_v = < b(y_i)\,B_i b(z_i)^{-1} (i \in K(v)) >$ of G.

Then $M_v = M_t$ if and only if $K(v) = K(t)$. Also $< M_v, M_t > \cong M_v * M_t$ if $K(v) \neq K(t)$ and $< M_v, E > \cong M_v * E$ for each v as subgroups of F. Similar considerations apply to the N_v in G.

By the GIC it follows that $M_v \cong N_v$ by the obvious map. Now we put

$$Y = < a(y_i)^{-1} A_i\ a(z_i)\,(i \in I) > * E \subset F$$

$$Z = < b(y_i)\ B_i\ b(z_i)^{-1} (i \in I) > * E \subset G.$$

Hence $Y \cong Z$ under the map $E \to E$ and $M_v \to N_v$, i.e.

$$a(y_i)^{-1}\ A_i\ a(z_i) \to b(y_i)\ B_i\ b(z_i)^{-1} \text{ for each } i.$$

Now setting

$$\overline{E} = < S, a(v),\ b(v)\,(v \in V);\ D,$$
$$a(y_i)^{-1}\ A_i\ a(z_i) = b(y_i)\ B_i\ b(z_i)^{-1} (i \in I) >$$

we see immediately that $\overline{E} \cong (F * G;\ Y = Z)$. The map

$$S \to S,\ p(v) \to a(v)b(v)$$

embeds E^* in $\overline{E}$. To see this observe that the relations of $\overline{E}$ are exactly the relations of E^* under this map so it is certainly a homomorphism. Moreover, if a word $W(S,p(v)) \to W(S,a(v)b(v)) = 1$ in $\overline{E}$, W must also equal 1 in the group T obtained from $\overline{E}$ by setting $b(v) = 1$. But clearly $T \cong E^*$ via the map $S \to S$, $a(v) \to p(v)$. Since the composite of all these maps is the identity on E^*, $W(S,p(v)) = 1$ in E^*. Hence E is embedded in E^* and $\bar{E}$ by $S \to S$ for $E \subset Y = Z$.

Now to prove the theorem we may assume W is freely reduced in E^* and

$$W \equiv S_1\, p(v_1)^{\alpha_1}\, S_2\, p(v_2)^{\alpha_2} \ldots S_n\, p(v_n)^{\alpha_n}\, S_{n+1}$$

where $n \geq 1$ and all $\alpha_i \neq 0$ and $S_i \in E$. Since $W = 1$ in E^* we must have for its embedded image in $\bar{E}$

$$W \equiv S_1\,(a(v_1)\,b(v_1))^{\alpha_1}\, S_2 \ldots S_n\,(a(v_n)\,b(v_n))^{\alpha_n}\, S_{n+1} = 1 \text{ in } \bar{E}.$$

Notice that $a(v) \in F\setminus Y$, $b(v) \in G\setminus Z$ and $S_i \in Y = Z$ since $\bar{E} = (F * G;\ Y = Z)$. By hypothesis $n \geq 1$. From Lemma 1 we conclude that for some i one of the following holds:

(1) $\alpha_i < 0$ and $\alpha_{i+1} > 0$ and $a(v_i)^{-1}\, S_{i+1}\, a(v_{i+1}) \in Y$

(2) $\alpha_i > 0$ and $\alpha_{i+1} < 0$ and $b(v_i)\, S_{i+1}\, b(v_{i+1})^{-1} \in Z$

(3) $\alpha_i < 0$ and $\alpha_{i+1} \leq 0$ and $a(v_i)^{-1}\, S_{i+1} \in Y$

(4) $\alpha_i > 0$ and $\alpha_{i+1} \geq 0$ and $b(v_i)\, S_{i+1} \in Z$.

But cases (3) and (4) are clearly impossible. So consider case (1). It follows from notation and free products that $a(v_i)^{-1}\, S_{i+1}\, a(v_{i+1}) \in M_v$ for some v, since $Y \cong E * R$ where R is the free product of the distinct M_v and such a word must have free product normal form length 1 (cancelling a(v) pairs must come from generators of the same M_v by our notation). Hence $S_{i+1} = w(A_\ell)$ in E for $w(A_\ell) \in A(v)$. Indeed, we have

$$a(v_i)^{-1} S_{i+1}\, a(v_{i+1}) = \prod_{j=1}^{m} a(t_j)^{-1}\, A_{\ell_j}^{\varepsilon_j}\, a(t_{j+1})$$

and

$$w(A_\ell) = \prod_{j=1}^{m} A_{\ell_j}^{\varepsilon_j} \quad \text{where } \varepsilon_j = \pm 1$$

and

$$a(v_i) = a(t_1) \text{ and } a(t_{m+1}) = a(v_{i+1})$$

and where the $a(t_j)^{-1} A_{\ell_j}^{\varepsilon_j} a(t_{j+1})$ are generators of M_v or their inverses.

Hence W contains a subword equal to $b(v_i)^{-1} a(v_i)^{-1} w(A_\ell) a(v_{i+1}) b(v_{i+1})$ which as a word of E^* has the form $p(v_i)^{-1} w(A_\ell) p(v_{i+1})$. But in $\bar{E}$ we have

$$a(v_i)^{-1} S_{i+1} a(v_{i+1}) = a(v_i)^{-1} w(A_\ell) a(v_{i+1}) = b(v_i) w(B_\ell) b(v_{i+1})^{-1}$$

where $w(A_\ell) \in A(v)$ and $w(B_\ell) \in B(v)$. Moreover $p(v_i)^{-1} w(A_\ell) p(v_{i+1}) = w(B_\ell)$ and $w(A_\ell) p(v_{i+1})$ produces $p(v_i) w(B_\ell)$ by use of the relations corresponding to the sequence of $a(t_j)$'s. This completes the proof in the situation arising in case (1) of the theorem.

Similarly, case (2) can be handled by the dual proof.||

If E^* and E are as above and the GIC holds, we say that E^* is a *Britton extension* of E (with respect to the stable letters P).

COROLLARY 2: *Let* E^* *be a Britton extension of* E. *Let* n *be a positive integer. Then* E^* *has elements of order* n *if and only if* E *has elements of order* n.

Proof: The claim is true for $\bar{E}$ (and hence E^*) by properties of free products with amalgamation, see [30].||

A group $G = E_n \geq E_{n-1} \geq \cdots \geq E_o = E$ is called a *Britton tower* over E if each E_i is a Britton extension of E_{i-1}.

COROLLARY 3: *Let* G *be a Britton tower over* E. *Let* n *be a positive integer. Then* G *has elements of order* n *if and only if* E *has elements of order* n.

Proof: By induction and Corollary 2.||

In the special case

$$E^* = (S,P;\ D,\ p(y_i)^{-1} A_i\ p(y_i) = B_i (i \in I))$$

we call E^* a *strong Britton extension* of E. Here, $p(v) \approx p(w)$ if and only if $p(v) = p(w)$. Then the GIC becomes the *isomorphism condition* of Britton (see [16] or [44] Chapter 12). Also $Tp(v)$ produces $p(w)U$ implies $p(v) = p(w)$. Hence Theorem A just becomes *Britton's Lemma.* A consequence of E being embedded in E^* is a well known theorem of Higman, Neumann and Neumann (see [26] or [44] Chapter 11).

LEMMA 4: *If* E^* *is a strong Britton extension of* E *with respect to the stable letters* P, *then the letters of* P *freely generate a free subgroup of* E^*

Proof: Let F be the free group on the letters P. Define a map $\phi: E^* \to F$ by $p(v) \to p(v)$ and $S \to 1$. Then, in view of the relations of E^*, ϕ is a homomorphism of E^* onto F. Hence any relation among the letters of P in E^* would also be a relation in F.||

Returning now to the general case in which E^* is a Britton extension of E, we indicate some ways in which Theorem A is applied. Let W be a word of E^*. Even though we may have $W \neq 1$ in E^*, the conclusion (II) of Theorem A may hold: say W contains a subword of the form (1) $p(y)^{-1}\, C\, p(z)$ where C is equal in E to a word $w(A_i) \in A(y)$ and $w(A_i)\, p(z)$ produces $p(y)\, w(B_i)$. Then W is seen to be equal in E^* to a word with two fewer p-symbols:

$$W \equiv U_1\, p(y)^{-1}\, C\, p(z) U_2 = U_1\, p(y)^{-1}\, w(A_i)\, p(z) U_2$$

$$= U_1\, p(y)^{-1}\, p(y)\, w(B_i) U_2 = U_1\, w(B_i) U_2.$$

A similar situation arises in the dual case (2). This process of replacing W by a word $U = W$ in E^* with two fewer P-symbols is called *pinching* the P-symbols out of W. In a particular pinching the P-symbols must be *equivalent,* i.e. $p(y) \approx p(z)$. A word W of E^* is called *P-free* if it involves no P-symbols. W is called *P-reduced* if W is not equal in E^* to a word with fewer P-symbols.

LEMMA 5: *Let* W *be a word of* E^*. *Then* W *is* P-*reduced if and only if no* P-*symbols can be pinched out of* W.

Proof: ($\rightarrow$) is obvious. To prove ($\leftarrow$) we suppose $W = U$ in E^* where U contains fewer P-symbols and is P-reduced. If U is P-free, then from $WU^{-1} = 1$ in E^* the conclusion follows from Theorem A. Suppose U is not P-free and consider $WU^{-1} = 1$ in E^*. By Theorem A, we must be able to remove all P-symbols from the product WU^{-1} by successive pinchings of equivalent P-symbols. Since no P-symbols can be pinched out of either W or U^{-1}, the P-symbols of U^{-1} must be pinched out with corresponding P-symbols of W. Hence U and W have the same number of P-symbols, a contradiction.||

Let W be a word of E^*. By the *P-projection* of W we mean the word in P-symbols obtained by erasing all of the letters in S occuring in W. Two words U and V of E^* are said to be *P-parallel* if their P-projections have the same number of P-symbols and the corresponding P-symbols are equivalent. We note from the proof of Lemma 5 that:

LEMMA 6: *Let* U *and* V *be* P-*reduced words of* E^*. *If* $U = V$ *in* E^*, *then* U *and* V *are* P-*parallel.*||

(See [18] for a more formal proof.)

LEMMA 7: *Let* S_1, S_2 *be words of* E *and let* V *be a word on the* P-*symbols alone. Then* $S_1 = S_2 V$ *in* E^* *if and only if* $V = 1$ *in* E^* *and* $S_1 = S_2$ *in* E. *If* E^* *is a strong Britton extension of* E, $V = 1$ *in* E^* *implies* V *is freely equal to 1.*

Proof: ($\leftarrow$) is obvious. ($\rightarrow$) Assume $S_1 = S_2 V$. Then $1 = S_1^{-1} S_2 V$ in E^*. By Theorem A we must be able to successively pinch all of the P-symbols out of V. Since V contains only P-symbols, the $w(A_i)$ and $w(B_i)$ in the conclusion (II) of Theorem A are equal to 1 in E. Thus after pinching out the P-symbols in V we are left with the equation $1 = S_1^{-1} S_2$. Hence $V = 1$ in E^* and $S_1 = S_2$ in E since E is embedded in E^* as claimed. The final remark follows from Lemma 4.||

Let us examine the $\bar{E}$ in the proof of Theorem A more closely. By using Tietze transformations we have the following:

$$\overline{E} = \langle S, a(v), b(v)\ (v \in V);\ D, a(y_i)^{-1} A_i\, a(z_i) = b(y_i)\, B_i\, b(z_i)^{-1}\ (i \in I) \rangle$$

$$\cong \langle S, a(v), b(v), p(v)\ (v \in V);\ D,\ p(v) = a(v)\, b(v),\ p(y_i)^{-1} A_i\, p(z_i) = B_i \rangle$$

$$\cong \langle S, a(v), p(v)\ (v \in V);\ D,\ p(y_i)^{-1} A_i\, p(z_i) = B_i \rangle$$

$$\cong E^* * \langle a(v)\ (v \in V) \rangle.$$

Similarly, one can show

$$\overline{E} \cong E^* * \langle b(v)\ (v \in V) \rangle.$$

Moreover, the isomorphisms involved are the natural ones.

We record these calculations in the following:

LEMMA 8: *Let* E^* *be a Britton extension of* E *with respect to the stable letters* $p(v)$ $(v \in V)$. *For* $\overline{E}$ *as in the proof of Theorem A, we have*

$$\overline{E} \cong E^* * \langle a(v)\ (v \in V) \rangle \ \|$$

B. COLLINS' LEMMA GENERALIZED

Suppose $G = (A_1 * A_2;\ H_1 = H_2, \phi)$. An element W of G is called *cyclically reduced* if W has normal form $XY_1 \ldots Y_n$ where $X \in H_1 = H_2$ and Y_1 and Y_n are not in the same factor unless $n = 1$. Clearly if $W = Z_1 \ldots Z_n$ where $n \geq 2$ and Z_i, Z_{i+1} are in distinct factors, then W is cyclically reduced if and only if Z_1 and Z_n are in distinct factors; for if W is written in normal form as $XY_1 \ldots Y_n$, then Y_i is in the same factor as Z_i. Now we consider Solitar's Theorem (for a proof see [30], Theorem 4.6 p. 212).

Solitar's Theorem: Let $G = (A_1 * A_2;\ H_1 = H_2, \phi)$. Then every element of G is conjugate to a cyclically reduced element of G. Moreover, suppose that W is a cyclically reduced element of G. Then:

(i) If W is conjugate to an element X in $H_1 = H_2$, then W is in some factor and there is a sequence $X, X_1, \ldots, X_t, W$ where X_i is in $H_1 = H_2$ and consecutive terms of the sequence are conjugate in a factor.

(ii) If W is conjugate to an element W' in some factor, but not in a conjugate of $H_1 = H_2$, then W and W′ are in the same factor and are conjugate in that factor.

(iii) If W is conjugate to an element $U_1 \dots U_n$ where $n \geq 2$ and U_i, U_{i+1} as well as U_1, U_n are in distinct factors, then W can be obtained by cyclically permuting $U_1 \dots U_n$ and then conjugating by an element of $H_1 = H_2$.

REMARK: Under the hypothesis of case (iii) above, $U_1 \dots U_n$ is cyclically reduced and W has length n.

Next we derive the analogue of case (iii) of Solitar's Theorem for Britton extensions. Suppose E^* is a Britton extension of E with respect to the stable letters P. A word W of E^* is called *P-cyclically reduced* if no cyclic permutation of W is equal to a word with fewer P-symbols (i.e. every cyclic permutation of W is P-reduced). Clearly, every word of E^* is conjugate to a P-cyclically reduced word of E^*. The following is a mild generalization of [18] General Lemma 3, p. 123:

Collins' Lemma: Let E^* be a Britton extension of E with respect to the stable letters P. Let U, $V \in E^*$ be P-cyclically reduced words and suppose U ends in $p(y)^{\varepsilon}$. Then $U \underset{E^*}{\sim} V$ if and only if there is a cyclic permutation V_0 of V such that

(i) U and V_0 are P-parallel.
(ii) V_0 ends in $p(z)^{\varepsilon}$ where $p(y) \approx p(z)$.
(iii) There is a word $S \in E$ such that

$U = S^{-1} V_0 S$ where for $\varepsilon = -1$ $S^{-1} \underset{E}{=} w(A_i) \in A(y)$ and $w(A_i)\, p(z)$ produces $p(y)\, w(B_i)$, while for $\varepsilon = 1$, $S^{-1} \underset{E}{=} w(B_i) \in B(y)$ and again $w(A_i)\, p(z)$ produces $p(y)\, w(B_i)$.

REMARK: There is of course a dual form of this lemma for the case U and V_0 begin with a P-symbol. We also refer to this dual formulation as *Collins' Lemma* and use whichever form is more convenient under the circumstances. Notice that Collins' Lemma implies that a P-cyclically reduced word of E^* is not conjugate to a word with fewer p-symbols.

Proof of Collins' Lemma: For the first part of the proof assume only that U, V are P-cyclically reduced words of E^*, V not P-free, and that the number of P-symbols of U is less than or equal to the number of P-symbols of V.

Assume that $U = p(y)^{-1} S_1^{-1} V S_1 p(y)$ in E^* where $S_1 \in E$. Then by the hypothesis on U and V either (1) the initial $p(y)^{-1}$ pinches with the first $p(z)^{\varepsilon}$ in V or (2) the terminal p(y) pinches with the last $p(z)^{\varepsilon}$ in V, where $\varepsilon = \pm 1$. Suppose case (1) occurs. Then $V \equiv S_2\, p(z)\, V_1$ where $S_2 \in E$ and for some $S_3 \in E$ we have $p(y)^{-1} S_1^{-1} S_2\, p(z) = S_3$. Moreover, $U = S_3\, V_1\, S_1\, p(y)$. But $S_1\, p(y) = S_2\, p(z)\, S_3^{-1}$ so that $U = S_3 V_1 S_2\, p(z) S_3^{-1}$. Observe that $V_1\, S_2\, p(z)$ is just a cyclic permutation of V. If case (2) occurs, a similar conclusion holds – namely that U is conjugate to a cyclic permutation of V by a word $S_3 \in E$.

More generally, suppose $U \underset{E^*}{\sim} V$. Then there exists $W \in E^*$ such that $U = W^{-1} VW$. Applying the above considerations, by induction on the number of P-symbols in W it follows that there is a cyclic permutation V_0 of V such that $U = S^{-1} V_0\, S$ for some P-free word $S \in E$. By Lemma 6, U and V_0 are P-parallel, so that U and V must have the same number of P-symbols. Thus, two P-cyclically reduced words which are conjugate in E^* have the same number of P-symbols.

Now assume the hypothesis of Collins' Lemma. The sufficiency is trivial. To prove the necessity, assume $U \underset{E^*}{\sim} V$. Then V is not P-free and U and V have the same number of P-symbols. Moreover, as above, there is a cyclic permutation V_0 of V such that $U = S^{-1} V_0\, S$ for some $S \in E$. Thus U and V_0 are P-parallel. Since U ends in $p(y)^{\varepsilon}$, S and V_0 can clearly be chosen so that V_0 ends in $p(z)^{\varepsilon}$. (Note that $p(y) \approx p(z)$ since U and V_0 are P-parallel.) From the equation $US^{-1} V_0^{-1} S = 1$, the conclusion (iii) follows by Theorem A since U and V_0 are P-reduced.||

By using the natural embedding of E^* in $\overline{E}$, one can readily deduce Collins' Lemma from Solitar's Theorem. However, the notational difficulties make the argument less transparent than the one given above which is a modification of Collins' proof. For a geometric proof see [33].

Convention on the use of Collins' Lemma:

Collins' Lemma will be used as a tool to help decide whether or not two words $U, V \in E^*$ are conjugate in E^* (possibly using an oracle). Assume that one can effectively P-reduce any word of E^*. Then one can effectively find for any $U, V \in E^*$ two P-cyclically reduced words conjugate to U and V respectively. So to decide conjugacy, one can assume U and V are P-cyclically reduced. Moreover, by Collins' Lemma, one can suppose that U and V are *P-circum-parallel,* i.e. a cyclic permutation of one is P-parallel to the other. If U (and hence V) is P-free Collins' Lemma does not apply. However, for U not P-free we use Collins' Lemma in the following way: Pick a particular permutation U_1 of U which ends in a $p(y)^{\varepsilon}$. Let V_1 be any permutation of V ending in $p(z)^{\varepsilon}$ such that U_1 and V_1 are P-parallel. Then we show how to decide, *for that particular permutation,* whether U_1 and V_1 are conjugate by a word $S \in E$ as in condition (iii) of Collins' Lemma. Since the number of such permutations is bounded by the number of P-symbols in V, this means we can decide whether $U \underset{E^*}{\sim} V$. Note that we may pick U_1 in advance, so long as U_1 ends in a $p(y)^{\varepsilon}$.

Our usual practice will be to then show how to decide for a particular V_1 as above whether or not there is an $S \in E$ as in condition (iii) such that $U_1 = S^{-1} V_1 S$. This implicitly defines an algorithm to decide whether or not $U \underset{E^*}{\sim} V$, so we omit the instruction saying that our procedure is to be applied to each of the remaining allowed permutations of V.

C. FURTHER PROPERTIES OF BRITTON EXTENSIONS

Let A_1 and A_2 be finitely presented groups. Baumslag [2] has shown that $G = (A_1 * A_2; H_1 = H_2, \phi)$ is f.p. if and only if $H_1 = H_2$ is finitely generated (f.g.). Now consider a Britton extension E^* of a f.p. group E with respect to a finite set of stable letters P. Then E^* will be f.p. if and only if each of the subgroups A(y) (see the definition of the GIC) is finitely generated. Indeed, since $\bar{E} \cong E^* * < a(v)\ (v \in V) >$ where V is finite, we have $\bar{E}$ f.p. $\leftrightarrow$ E^* f.p., so this follows from the above result of Baumslag.

Suppose E^* as above is f.p. Then from properties of ordinary free products and the isomorphism $\bar{E} = E^* * < a(v)\ (v \in V) >$ we conclude that the word problem for $\bar{E}$ has the same degree as the word problem for E^*. Moreover, the conjugacy problem for $\bar{E}$ has the same degree as the conjugacy problem for E^*.

Suppose now that the underlying group E is free of finite rank. Then $\bar{E}$ is the free product with amalgamation of two free groups. Since free groups have solvable generalized word problems, from [30] Lemma 4.9 p. 272 it follows that $\bar{E}$ has solvable word problem. Consequently:

LEMMA 9: *If* E^* *is a f.p. Britton extension of a free group* E, *then* E^* *has solvable word problem.* ||

For more information concerning the word problem for free products with amalgamation and for strong Britton extensions see Clapham [17], particularly pp. 644-646.

We will frequently use the following facts about free and direct products: Let $G = A * B$ or $G = A \times B$. Then the degree of the word problem (respectively conjugacy problem) for G is equal to the join of the degrees of the word problems (respectively conjugacy problems) for A and B. This is easily verified by well-known structural results for free and direct products. (See [30] in this connection).

CHAPTER III

UNSOLVABILITY RESULTS FOR RESIDUALLY FINITE GROUPS

A. A F.P. RESIDUALLY FINITE GROUP WITH UNSOLVABLE CONJUGACY PROBLEM

Let $H = \langle s_1,\ldots,s_n;\ R_1,\ldots,R_m \rangle$ be a finitely presented group with unsolvable word problem. Put $F = \langle q,s_1,\ldots,s_n \rangle$, a free group of rank $n+1$ on the listed generators. Finally, define the f.p. group G as follows:

Generators:

$$q,\ s_1,\ldots,s_n,\ t_1,\ldots,t_m,\ d_1,\ldots,d_n$$

Relations:

$$\left.\begin{aligned} t_i^{-1}\, q\, t_i &= q\, R_i \\ t_i^{-1}\, s_\alpha\, t_i &= s_\alpha \end{aligned}\right\} \qquad \begin{aligned} &1 \leq i \leq m \\ &1 \leq \alpha \leq n \end{aligned}$$

$$\left.\begin{aligned} d_\alpha^{-1}\, q\, d_\alpha &= s_\alpha^{-1}\, q\, s_\alpha \\ d_\alpha^{-1}\, s_\beta\, d_\alpha &= s_\beta \end{aligned}\right\} \qquad \begin{aligned} &1 \leq \alpha \leq n \\ &1 \leq \beta \leq n \end{aligned}$$

where the R_i are the words in s_i which appear as the given defining relations for H.

We will show that (1) G has unsolvable conjugacy problem and (2) G is residually finite. The proof will be given in some detail since this is the first example of a basic type of construction which will reappear in Chapter IV.

LEMMA 1: G *is a strong Britton extension of* F *with respect to the stable letters* $\{t_1,\ldots,t_m,d_1,\ldots,d_n\}$.

Proof: Each of the Britton subgroups in question is F itself. For example, for fixed i, the subgroups involved with t_i are generated by $\{q, s_1, \ldots, s_n\}$ and $\{qR_i, s_1, \ldots, s_n\}$. Since each of these sets consists of $n+1$ elements of F which generate F which has rank $n+1$, they freely generate F. Consequently, the map $q \to qR_i$, $s_\alpha \to s_\alpha$ generates an automorphism of F as was to be verified. The same argument applies to the subgroups associated with each d_α.||

Now rewrite the relations given for G which involve q as follows:

$$q^{-1} \quad t_i \quad q = t_i R_i^{-1} \qquad (1 \leq i \leq m)$$

$$q^{-1} \, d_\alpha s_\alpha^{-1} \quad q = d_\alpha s_\alpha^{-1} \qquad (1 \leq \alpha \leq n).$$

Put $K = \langle t_1, \ldots, t_m, d_1, \ldots, d_n \rangle \times \langle s_1, \ldots, s_n \rangle$, the direct product of the two listed free groups.

LEMMA 2: G *is a strong Britton extension of* K *with respect to the stable letter* q.

Proof: In K the subgroups generated by

$$t_1, \ldots, t_m, d_1 s_1^{-1}, \ldots, d_n s_n^{-1}$$

and $t_1 R_1, \ldots, t_m R_m, d_1 s_1^{-1}, \ldots, d_n s_n^{-1}$ are both free and are isomorphic under the map $t_i \to t_i R_i$ and $d_\alpha s_\alpha^{-1} \to d_\alpha s_\alpha^{-1}$ as was to be verified.||

From Lemma 1 and Lemma II-9 it follows that:

LEMMA 3: G *has solvable word problem.*||

NOTATION: Let X, Y, Z be variables for words on the s_α. For $X \equiv X(s_\alpha)$ a word on the s_α, $X(d_\alpha)$ is the word on the d_α obtained by replacing each s_α by the corresponding d_α. Let T be the subgroup of G generated by the t_i and d_α.

Some sample calculations in G may help in understanding the construction. Note that any word in T commutes with any word on the s_α.

$$(d_1^{-1}d_2^{-1})\ q\ (d_1^{-1}d_2^{-1})^{-1} = d_1^{-1}d_2^{-1}\ q\ d_2d_1$$

$$= d_1^{-1}s_2^{-1}\ q\ s_2d_1$$

$$= s_2^{-1}d_1^{-1}\ q\ d_1s_2$$

$$= s_2^{-1}s_1^{-1}\ q\ s_1s_2$$

$$= (s_1s_2)^{-1}\ q\ (s_1s_2).$$

More generally, for Z a word on the s_α,

$$Z(d_\alpha^{-1})\ q\ Z(d_\alpha^{-1})^{-1} = Z^{-1}\ q\ Z.$$

Hence, $Z(d_\alpha^{-1})\ Z\ q\ Z(d_\alpha^{-1})^{-1} = q\ Z$ and $Z(d_\alpha^{-1})^{-1}\ q\ Z(d_\alpha^{-1}) = Z\ q\ Z^{-1}$.

Using these one calculates that for $W \equiv Z(d_\alpha^{-1})t_i\ Z(d_\alpha^{-1})^{-1}$:

$$W^{-1}\ q\ W = Z(d_\alpha^{-1})t_i^{-1}\ \ Z\ q\ Z^{-1}t_i\ Z(d_\alpha^{-1})^{-1}$$

$$= Z(d_\alpha^{-1})\ Z\ q\ R_i\ Z^{-1}\ Z(d_\alpha^{-1})^{-1}$$

$$= q\ Z\ R_i\ Z^{-1}.$$

Notice that W is a word of T. The following result is fundamental:

LEMMA 4:

(1) $(\exists\, W \in T)\ (W^{-1}X_1qY_1W \underset{G}{=} X_2qY_2) \leftrightarrow X_1Y_1 \underset{H}{=} X_2Y_2.$

(2) $X_1qY_1 \underset{G}{\sim} X_2qY_2 \leftrightarrow X_1Y_1 \underset{H}{\sim} X_2Y_2.$

Proof: First consider (1). $(\rightarrow)$ of (1) is clear from the defining relations for G by induction on the length of $W \in T$. To prove $(\leftarrow)$ of (1), assume that $X_1Y_1 = X_2Y_2$ in H. Then $X_2Y_2 = (\prod_{i=1}^{k} Z_i\ R_{\ell_i}^{\varepsilon_i}\ Z_i^{-1})X_1Y_1$ in F. Put $\theta_i = Z_i(d_\alpha^{-1})t_{\ell_i}^{\varepsilon_i}\ Z(d_\alpha^{-1})^{-1}$ and let $W^* = \theta_k\ldots\theta_2\theta_1$. Then $W^* \in T$ and $W^{*-1}s_\alpha\ W^* = s_\alpha$. A calculation now gives $W^{*-1}\ q\ X_1Y_1\ W^* = q\ X_2Y_2$.

Finally, put $W = X_1(d_\alpha^{-1})^{-1} W^* X_2(d_\alpha^{-1}) \in T$. Then one checks that $W^{-1} X_1 q Y_1 W = X_2 q Y_2$, so W is the desired conjugating element.

Now consider (2). To prove (←) of (2), assume $X_1Y_1 \underset{H}{\sim} X_2Y_2$. Then there is a word X_0 on the s_α such that $X_0^{-1}X_1Y_1X_0 \underset{H}{=} X_2Y_2$. By (1), $\exists W \in T$ such that

$$W^{-1}X_0^{-1}X_1 q Y_1X_0W = X_2 q Y_2.$$

Hence $X_1 q Y_1 \underset{G}{\sim} X_2 q Y_2$ (by the word X_0W) as claimed. Finally to prove (→) of (2), assume $X_1 q Y_1 \underset{G}{\sim} X_2 q Y_2$. Observe that the presentation obtained from that for G by adding the relations $q = 1$, $t_i = 1$ $(1 \le i \le m)$, and $d_\alpha = 1$ $(1 \le \alpha \le n)$ is just another presentation for H. Hence there is an epimorphism ϕ from G onto H which sends $s_\alpha \to s_\alpha$ and all other letters to 1. Now $\phi(X_1 q Y_1) = X_1Y_1$ and $\phi(X_2 q Y_2) = X_2Y_2$. Consequently, $W^{-1}X_1 q Y_1W \underset{G}{=} X_2 q Y_2$ implies $\phi(W)^{-1}X_1Y_1 \phi(W) \underset{H}{=} X_2Y_2$ as claimed.||

COROLLARY 5: G *has unsolvable conjugacy problem.*

Proof: Since H has unsolvable word problem, H also has unsolvable conjugacy problem. By Lemma 4, G also has unsolvable conjugacy problem.||

Now T is the subgroup of G generated by the t_i and the d_α. In view of Lemma 2, T is a free group. (Alternatively, this follows from Lemma 1 and Lemma II-4). Recalling the proof of Lemma 1 (or by simply looking at the relations for G), one has $F \lhd G$ and $G = FT$. Moreover, $F \cap T = \{1\}$. For suppose $U_1, U_2 \in F$ and $V \in T$. Then $U_1 = U_2V$ implies $U_1 \underset{F}{=} U_2$ and $V \underset{T}{=} 1$ because of Lemmas 1 and II-7. Hence:

LEMMA 6: G *is the split extension of the f.g. free group* F *by the f.g. free group* T.||

In [3] Baumslag has shown that the automorphism group of a f.g. residually finite group is itself residually finite. (see also [30] p. 414). By imitating that argument, we show the following general result:

THEOREM 7: *Suppose that* $1 \to A \to E \overset{\pi}{\to} B \to 1$ *is an exact sequence of groups (i.e.* E *is an extension of* A *by* B*) where* A *and* B *are residually finite and* A *is f.g. Assume that any one of the following conditions holds:*

(1) A *has trivial* center.

(2) *The sequence splits (i.e.* E *is a split extension or semi-direct product of* A *by* B).

(3) B *is free or* A *is non-abelian free.*

Then E *is residually finite.*

REMARK: This result under hypothesis (2) is stated in Mal'cev [31].

Proof: Denote the centralizer in a group K of a group J by $C_K(J)$ and by $Z(K) = C_K(K)$ the center of a group K. Let $1 \neq W \in E$. We must show there is a map $\phi : E \to M$ (a finite group) such that $\phi(W) \neq 1$ in M. Condition (3) implies either condition (1) or condition (2), so we need only verify that E is residually finite assuming (1) or (2). A will be viewed as a subgroup of E.

CASE (i): $W \notin A$. Then under the natural epimorphism $\pi : E \to B = E/A$ we have $1 \neq \pi(W) \in B$. Since B is residually finite, there is an epimorphism $\Psi : B \to M$ where M is a finite group such that $1 \neq \Psi(\pi(W)) \in M$ as required.

CASE (ii): Assume condition (2) and $W \in A$. Since A is residually finite, A has a normal subgroup N of finite index n such that $W \notin N$. Let N^* be the intersection of all normal subgroups of A having index n. According to M. Hall [24], N^* has finite index (since A is f.g.) and is characteristic in A. Hence $N^* \lhd E$ and there is a natural epimorphism $\delta : E \to C$ with kernel N^*. By condition (2), $E = A \cdot B$, $A \lhd E$, and $A \cap B = \{1\}$. Now applying δ, it follows that C is the semi-direct product $(A/N^*) \cdot B$ with action induced from the semi-direct product $E = A \cdot B$.

Now $(A/N^*) \lhd C$, $B \cap (A/N^*) = \{1\}$, and $1 \neq \delta(W) \in A/N^* \subseteq C$. Recall that A/N^* is finite. Put $Q = C_B(A/N^*) \subseteq C$, i.e. Q is the set of all $U \in B$ such that $U^{-1}VU = V$ for every $V \in A/N^*$. Then Q is in fact a normal subgroup of B. Moreover, for each $V \in A/N^*$ it follows that $V^{-1}QV = Q$ so $Q \lhd C$. Now since $Q \subseteq B$ and $Q \lhd C$ there is a natural epimorphism $\gamma : C \to M$ where M is the semi-direct product $(B/Q) \cdot (A/N^*)$. Now each element of B/Q acts as a non-trivial automorphism on the finite group A/N^*. Hence M is finite. Moreover, $1 \neq \gamma(\delta(W)) \in A/N^* \subseteq M$ as required.

CASE (iii): Assume condition (1) and $W \in A$. Since $W \notin Z(A) = 1$ (by condition (1)), $\exists X \in A$ such that $1 \neq X$ and $1 \neq [W,X]$. Hence $\exists K \lhd A$ such that $[W,X] \notin K$ and A/K is finite (hence also $W \notin K$ and $X \notin K$). Now $W \notin Z(A/K)$, since $1 \neq [W,X]$ in A/K; also $W \in A/K$ and $X \in A/K$. Since A is f.g., as in case (ii), $\exists K^* \subseteq K \subset A$ such that K^* is characteristic in A of finite index. Thus $W \notin Z(A/K^*)$ and $K^* \lhd E$ (since K^* is characteristic in the normal subgroup A). Put $J = E/K^*$, $L = A/K^* \lhd J$. Now L is finite, $W \in L$, but $W \notin Z(L)$. Consider $C_J(L)$.

Now (a) $W \notin C_J(L)$ since $W \notin Z(L) = C_J(L) \cap L$ and $W \in L$.

(b) $C_J(L) \lhd J$ since $L \lhd J$.

(c) $J/C_J(L)$ is finite. For suppose $U \notin C_J(L)$.

Then U induces by conjugation a non-trivial permutation on $L \lhd J$ (otherwise $U \in C_J(L)$). Since L is finite, there are only finitely many such non-trivial permutations, hence only finitely many equivalence classes mod $C_J(L)$, i.e. $J/C_J(L)$ is finite. Therefore $1 \neq W \in J/C_J(L)$ and $J/C_J(L)$ is a finite quotient of E as required.||

Since free groups are residually finite (see [30] p. 116 or 414) it follows from Lemma 6 and Theorem 7 that

LEMMA 8: G *is residually finite.*||

Combining Lemmas 6 and 8 with Corollary 5 we have shown:

THEOREM 9: *There exists a finitely presented, residually finite group* G *with unsolvable conjugacy problem. Moreover,* G *is the split extension of one free group by another.*||

As a result of Lemma 1, Corollary 5, and the discussion of Section C of Chapter II, it follows that:

THEOREM 10: *The free product of two free groups with finitely generated amalgamation can have unsolvable conjugacy problem. Further, the finitely presented strong Britton extension of a free group can have unsolvable conjugacy problem.*||

REMARKS: The proof of Theorem 9 can be given in an elementary way without using results of Chapter II. In Chapter IV arbitrary r.e. degree analogues of Theorems 9 and 10 will be obtained.

In view of Corollary 5 and Lemma 6, one observes the conjugacy problem for G is unsolvable because of the action of T on the free group F qua automorphisms. The next natural question is "does the automorphism group of a free group have unsolvable problems?" Section B below gives an affirmative answer.

B. UNSOLVABLE PROBLEMS IN THE AUTOMORPHISM GROUP OF A FREE GROUP

Call a recursively enumerable (r.e.) set C *suitably given* if it is given in such a way that one of the known effective constructions [11], [17], [23] for a finitely presented group with word problem of the same degree as C applies (e.g. given as the Gödel number of a Turing machine which semi-computes the characteristic function of C). An r.e. degree D is *suitably given* when an r.e. set in D is suitably given. Our first goal is to prove the following:

THEOREM 11: *Let* D *be an* r.e. *degree of unsolvability. Let* F *be a free group of finite rank* ≥ 3, Φ *the automorphism group of* F. *Then there is a finitely generated subgroup* A_D *of* Φ *whose membership problem (generalized word problem) has degree* D. *Moreover, if* D *is suitably given, a finite set of generators for* A_D *can be effectively found.*

The group Φ is residually finite [30, p. 414] and has a solvable word problem. Several presentations are known for Φ [30, pp. 162-165].

Assume the r.e. degree D is suitably given. Then there is an effective construction yielding a finitely presented (f.p.) 2-generator group $H_2 = \langle s_1, s_2; Q_1, \ldots, Q_\ell \rangle$ with the word problem of degree D.[1] The f.p. group

$$H_n = H_2 * \langle s_3, \ldots, s_n; s_3, \ldots, s_n \rangle \ (n \geq 2)$$

is isomorphic to H_2 since $\langle s_3, \ldots, s_n; s_3, \ldots, s_n \rangle$ is a presentation of the trivial group. Write

$$H_n = \langle s_1, s_2, s_3, \ldots, s_n; R_1, \ldots, R_m \rangle$$

where $R_1, \ldots, R_m$ are the relations $Q_1, \ldots, Q_\ell$ together with the relations $s_3, \ldots, s_n$. Observe that H_n also has word problem of degree D. Put $F_{n+1} = \langle q, s_1, \ldots, s_n \rangle$, a free group of finite rank $n + 1 \geq 3$. Let Φ_{n+1} be the automorphism group of F_{n+1}. Finally, let G_{n+1} be the group G constructed in Section A above using H_n for H and F_{n+1} for F. The following is a restatement of Lemma 1:

LEMMA 12: *The map* $q \to q$, $s_\alpha \to s_\alpha$ *embeds* F_{n+1} *in* G_{n+1}. *Moreover, conjugation of* F_{n+1} *by* t_i *or* d_α *induces an automorphism of* F_{n+1}. ||

T is the subgroup of G_{n+1} generated by the t_i and d_α. Recall that conjugation by each t_i and d_α sends a set of free generators of F_{n+1}

---

1 [11], [17], and [23] give effective constructions for a f.p. group K with word problem of degree D. J.L. Britton (oral communication) has shown K can be effectively embedded in a 2-generator f.p. group with word problem of degree D (see [34]). Alternatively, one can show the Higman, Neumann, and Neumann 2-generator embedding preserves the degree of the word problem.

onto another set of free generators of F_{n+1}. Hence to each t_i and d_α there correspond automorphisms $\phi(t_i)$ and $\phi(d_\alpha)$ of F_{n+1} whose action on the generators of F_{n+1} is exactly the same as conjugation by t_i and d_α respectively. Let A be the subgroup of Φ_{n+1} finitely generated by the $\phi(t_i)$ and $\phi(d_\alpha)$. From the presentation of Φ_{n+1} given in [30], it is clear that one can effectively find the generators of A from the presentation for G_{n+1}.

LEMMA 13: *Let* $\Psi \in \Phi_{n+1}$ *be an arbitrary automorphism. Then*

$$\Psi \in A \leftrightarrow (1)\ \Psi(s_\alpha) = s_\alpha\ (1 \le \alpha \le n)$$

$$\textit{and} \quad (2)\ \exists\, X,Y \ \textit{such that}\ \Psi(q) = X\,q\,Y$$

$$\textit{and} \quad (3)\ \exists\, X,Y \ \textit{such that}\ \Psi(q) = X\,q\,Y \ \textit{and}$$

$$\exists\, W \in T \ \textit{such that}\ W^{-1}qW \underset{G_{n+1}}{=} XqY.$$

Moreover, there is an algorithm to determine of arbitrary $\Psi \in \Phi_{n+1}$ *whether or not (1) and (2) hold, while whether or not (3) holds is a problem of r.e. degree* D *of unsolvability.*

REMARK: (2) is redundant in view of (3). The lemma is stated in this manner for future reference.

Proof: By the definition of A, $\Psi \in A \leftrightarrow$ there exists $W \in T$ whose action on the generators of F_{n+1} by conjugation is the same as the action of Ψ on the generators of F_{n+1}. But for any $W \in T$ we have $W^{-1}s_\alpha W = s_\alpha$ and $\exists\, X,Y$ such that $W^{-1}qW \underset{G_{n+1}}{=} XqY$. Thus, $\Psi \in A \leftrightarrow$ (1) and (2) and (3) follows immediately. The presentation of Φ_{n+1} given in [30, pp. 162-165] specifies the generators of Φ_{n+1} by giving their action on the generators of F_{n+1}. Hence for any $\Psi \in \Phi_{n+1}$ we can effectively determine whether or not (1) and (2) hold by merely computing the action of Ψ on the generators of F_{n+1}.

On the other hand, by Lemma 4.1 with X_1, Y_1 the empty word and the fact that H_n has word problem of degree D it follows that: the problem of deciding for arbitrary X,Y whether or not $\exists W \epsilon T$ such that $W^{-1}qW \underset{G_{n+1}}{=} XqY$ has degree D. Hence whether or not (3) holds has degree D.||

COROLLARY 14: *The problem of deciding for arbitrary* X,Y *whether or not* $\exists W \epsilon T$ *such that* $W^{-1}qW = XqY$ *in* G_{n+1} *is a problem of r.e. degree* D *of unsolvability.*||

COROLLARY 15: *The membership problem (generalized word problem) for* A *in* Φ_{n+1} *has degree* D *of unsolvability.*||

This completes the proof of the Theorem 11. Note that while the subgroup T of G_{n+1} is in fact free, the induced subgroup A may not be free since a non-trivial word in T can commute with q. The proof of Theorem 1 given above uses a particular presentation of Φ_{n+1} given in [30]. However since all of the conclusions are algebraic invariants, any other finite presentation could have been used.

Turning next to the conjugacy problem, we will show the following:

THEOREM 16: *Assume* $n \geq 3$. *Then the subgroup* A *of* Φ_{n+1} *has conjugacy problem of degree* $\geq$ D.

REMARK: While A is certainly f.g., it is not too difficult to prove that A is generally not finitely related. Consequently, we do not know whether there is a f.p. subgroup of Φ_{n+1} with unsolvable conjugacy problem. Even worse, the status of the conjugacy problem for Φ_{n+1} itself is apparently unknown.

Proof of Theorem 16: Since $n \geq 3$, the generator s_3 of H_n is present. Moreover, $s_3 = 1$ in H_n. Let θ be the automorphism of F_{n+1} sending $s_\alpha \to s_\alpha$ and $q \to s_3^{-1} q s_3$. (of course θ is just $\phi(d_3)$). For any words Σ, Δ on the letters s_1, s_2, let $\gamma(\Sigma,\Delta)$ be the automorphism of F_{n+1}

sending $s_\alpha \to s_\alpha$ and $q \to \Sigma^{-1} s_3^{-1} \Sigma q \Delta s_3 \Delta^{-1}$. Since $s_3 = 1$ in H_n, it follows from Lemmas 13 and 4 that $\theta \in A$ and $\gamma(\Sigma, \Delta) \in A$.

Now let Ψ be any automorphism of F_{n+1} such that $\Psi(s_\alpha) = s_\alpha$ $(1 \leq \alpha \leq n)$ and $\exists\, X, Y$ such that $\Psi(q) = XqY$. (i.e. Ψ has properties (1) and (2) of Lemma 13). One easily checks that $\Psi \theta \Psi^{-1}$ sends $s_\alpha \to s_\alpha$ and $q \to X^{-1} s_3^{-1} XqYs_3Y^{-1}$. Now $\Psi \theta \Psi^{-1} \in A$, but $\Psi \in A$ if and only if $XY \underset{H_n}{=} 1$ by Lemmas 13 and 4.

But $\Psi \theta \Psi^{-1} = \gamma(\Sigma, \Delta)$ if and only if $\Sigma^{-1} s_3^{-1} \Sigma \underset{F_{n+1}}{=} X^{-1} s_3^{-1} X$ and $\Delta s_3 \Delta^{-1} \underset{F_{n+1}}{=} Y s_3 Y^{-1}$. Since Σ, Δ are words on s_1, s_2 and X, Y words on the s_α, by properties of free groups it follows that $\Psi \theta \Psi^{-1} = \gamma(\Sigma, \Delta)$ if and only if $X \underset{F_{n+1}}{=} s_3{}^k \Sigma$ and $Y \underset{F_{n+1}}{=} \Delta s_3{}^\ell$. Now $\Psi \in A$ if and only if $XY \underset{H_n}{=} 1$. Consequently $\exists\, \Psi \in A$ such that $\Psi \theta \Psi^{-1} = \gamma(\Sigma, \Delta)$ if and only if $s_3{}^k \Sigma \Delta s_3{}^\ell \underset{H_n}{=} 1$; i.e. since $s_3 \underset{H_n}{=} 1$ if and only if $\Sigma \Delta \underset{H_n}{=} 1$.

Hence θ is conjugate in A to $\gamma(\Sigma, \Delta)$ if and only if $\Sigma \Delta \underset{H_n}{=} 1$. By properties of H_n, the result follows. ||

C. UNSOLVABLE PROBLEMS IN DIRECT PRODUCTS OF FREE GROUPS

In [52] Mihailova has shown the direct product of two free groups has unsolvable generalized word problem. Her construction is given below in Theorem 17 and Lemma 18. We will then apply this construction to establish a variety of unsolvability results for the direct product of free groups and for certain unimodular groups. For notational convenience mappings will be written on the right (in this section only).

THEOREM 17: *Let* M *be a finitely presented group, and let* H *be any f.p. quotient of* M *having word problem of degree* D *of unsolvability. Then the group* $G = M \times M$ *has a finitely generated subgroup* L_H *such that the generalized word problem for* L_H *in* G *has degree* D *of unsolvability.*

Proof: By hypothesis, we can assume that H has the presentation $<s_1,\dots,s_n;\ R_1,\dots,R_m>$ where M has the presentation $<s_1,\dots,s_n;\ R_1,\dots,R_\ell>$ with $\ell \leq m$ (so H is obtained from M by adding the relations $R_{\ell+1},\dots,R_m$). An element of the direct product $G = M \times M$ will be viewed as an ordered pair (X,Y). Now define $a_i = (s_i,s_i) \in G$ $(1 \leq i \leq n)$ and $b_j = (1,R_j) \in G$ $(1 \leq j \leq m)$. Let $L = L_H$ be the subgroup of G finitely generated by the a_i $(1 \leq i \leq n)$ and the b_j $(1 \leq j \leq m)$. The theorem is an immediate consequence of the following lemma:

LEMMA 18: $(X,Y) \in L \leftrightarrow X \underset{H}{=} Y$.

Proof: Let ϕ: $M \to H$ be the canonical map. Put $K = H \times H$ and define the epimorphism ψ: $G \to K$ by $(X,Y)\psi = (X\phi, Y\phi)$. Clearly $b_i\psi \underset{K}{=} (1,1)$.

Assume $(X,Y) \in L$. Then (X,Y) is a product of the a_i, the b_i and their inverses. But the a_i are diagonal elements of G and $b_i\psi \underset{K}{=} (1,1)$. Hence $(X,Y)\psi$ is a diagonal element of $K = H \times H$. That is, $X \underset{H}{=} Y$ as claimed.

Assume, conversely, that $X \underset{H}{=} Y$. Then $XY^{-1} \underset{H}{=} 1$ so

$$XY^{-1} \underset{M}{=} \prod_{k=1}^{r} U_k(s_i)R_{j_k}^{\varepsilon_k}U_k(s_i)^{-1}$$

since H is a quotient of M.

Now observe that in $G = M \times M$:

$$U_k(a_i)b_{j_k}^{\varepsilon_k} U_k(a_i)^{-1} \underset{G}{=} (1, U_k(s_i)R_{j_k}^{\varepsilon_k} U_k(s_i)^{-1}).$$

Consequently,

$$\prod_{k=1}^{r} U_k(a_i)b_{j_k}^{\varepsilon_k} U_k(a_i)^{-1} \underset{G}{=} (1, \prod_{k=1}^{r} U_k(s_i)R_{j_k}^{\varepsilon_k} U_k(s_i)^{-1})$$

$$\underset{G}{=} (1, XY^{-1}) \in L.$$

But L contains all diagonal elements of G and $(X,X) \in L$ in particular. Since $(1,XY^{-1}) \in L$ it follows that $(X,Y) \in L$ as claimed.||

This completes the proof of Theorem 17.

Let A and B be finitely generated groups and suppose ϕ: $A \to B$ is a monomorphism. The map ϕ is a *recursive embedding* of A into B if the GWP for $A\phi$ in B is recursively solvable. The following lemma is immediate:

LEMMA 19: *Let* L *be a finitely generated subgroup of* A *whose GWP in* A *has degree* D *of unsolvability. Suppose* ϕ: $A \to B$ *is a recursive embedding of* A *into* B. *Then the GWP for* $L\phi$ *in* B *has degree* D *of unsolvability. Moreover, if* $\psi : B \to C$ *is a recursive embedding of* B *into* C, *then* $\phi\psi$: $A \to C$ *is a recursive embedding of* A *into* C.||

LEMMA 20: *Let* F_n *be a free group of rank* $n \geq 2$, *and let* A *be any finitely generated free group. Then* A *can be recursively embedded in* F_n *and* $A \times A$ *can be recursively embedded in* $F_n \times F_n$.

Proof: Let $f_1,\ldots,f_n$ be a set of free generators for F_n and let $a_1,\ldots,a_\ell$ be a set of free generators for A. The subgroup of F_n generated by $f_1, f_2^{-1}f_1f_2, \ldots, f_2^{-(\ell-1)}f_1f_2^{\ell-1}$ is a free group of rank ℓ freely generated by these words (they are a Nielsen basis for the subgroup, see [30], section

3.2). Hence the map ϕ: $A \to F_n$ defined by

$$a_i\phi = f_2^{-(i-1)} f_1 f_2^{i-1} \quad (1 \le i \le \ell)$$

is a recursive embedding of A into F_n, because the GWP for $A\phi$ in F_n is solvable. It follows immediately that ψ: $A \times A \to F_n \times F_n$ defined by $(x,y)\psi = (x\phi, y\phi)$ is a recursive embedding.||

THEOREM 21: *Let* F *be a finitely generated free group of rank at least* 2 *and let* D *be a recursively enumerable degree of unsolvability. Then* $G = F \times F$ *has a finitely generated subgroup* L_D *such that the generalized word problem for* L_D *in* G *has degree* D.

Proof: By the main result of [11], there exists a finitely presented group H with word problem of degree D. Now H is a quotient of a suitable finitely generated free group A. By Theorem 1, $A \times A$ has a finitely generated subgroup L with GWP in $A \times A$ of degree D. By Lemma 20, $A \times A$ can be recursively embedded in $F \times F$, say by the map ψ. By Lemma 19, $L_D = L\psi$ has GWP in $G = F \times F$ of degree D as claimed.||

REMARKS: The group $G = F \times F$ in Theorem 21 seems to be rather elementary. G has solvable word problem and solvable conjugacy problem. Further, G is residually free and hence residually finite. Note that the (ordinary) free product $F*F$ is again a free group and so has solvable GWP for finitely generated subgroups. Thus Theorem 21 indicates that direct products can be more complicated than free products. Further evidence of this is provided by the following:

THEOREM 22: *Let* F *be a free group of rank at least* 9 *and let* $G = F \times F$. *Then the problem to determine of an arbitrary finite set of words whether or not they generate* G *is recursively unsolvable.*

Proof: Let U be a finitely presented group on two generators having unsolvable word problem. In [41] (see also section V-C) a construction is given which yields a recursive class $\Omega = \{H_W, W \in U\}$ of finite presentations of groups H_W indexed by words of U such that $H_W \cong 1 \leftrightarrow W \underset{U}{=} 1$. Moreover, each presentation H_W has exactly 9 generators (= number of generators of U+7). Since F has rank at least 9, we can regard each H_W as a quotient of F. Let L_W be the corresponding subgroup of $G = F \times F$ constructed in the proof of Theorem 17.

Assume $L_W = G$. Then for every $X,Y \in F$ we have $(X,Y) \in L_W$. Thus, by Lemma 18, for every $X,Y \in F$ we have $X \underset{H_W}{=} Y$. Hence, $H_W \cong 1$ and so $W \underset{U}{=} 1$. Conversely, suppose $W \underset{U}{=} 1$ so that $H_W \cong 1$. Then, for every $X,Y \in F$, $X \underset{H_W}{=} Y$ and, by Lemma 18, $(X,Y) \in L_W$. That is $L_W = G$. Hence $L_W = G \leftrightarrow W \underset{U}{=} 1$. Now each L_W is given by a finite set of words which depend explicitly on the presentation of H_W. Since Ω is recursive, one can decide if a given finite set of words in G are the given generators of an L_W. But U has an unsolvable word problem, so one can not decide whether such a finite set of words generate G because $L_W = G \leftrightarrow W \underset{U}{=} 1$.||

THEOREM 23: *Let* F *be a free group of rank at least* 2. *Put* $G = F \times F$. *Then* G *has a finitely generated subgroup* L *such that: (1)* L *has unsolvable conjugacy problem and (2) the generalized word problem for* L *in* G *is unsolvable. Note that* G *and hence* L *are residually free groups.*

Proof: Since having unsolvable conjugacy problem is an algebraic property of a group, from Lemmas 19 and 20 it follows that it suffices to prove the theorem for the case F has rank 3.

Let H be a 2-generator group with unsolvable word problem. Then we may take $<s_1,s_2s_3;\ R_1,\ldots,R_{m-1},s_3>$ to be our presentation for H (of course $s_3 \underset{H}{=} 1$). Let $F = <s_1,s_2,s_3>$, a free group of rank 3. Finally, take L to be the subgroup of $G = F \times F$ constructed in the proof of Theorem 17. Now L is finitely generated and has unsolvable generalized word problem by Theorem 17.

Let W be any word of F. Then $(s_3,s_3) \in L$ and, by Lemma 18, $(s_3,W^{-1}s_3W) \in L$ since $s_3 \underset{H}{=} W^{-1}s_3W \underset{H}{=} 1$.

LEMMA 24: (s_3,s_3) *is conjugate to* $(s_3,W^{-1}s_3W)$ *in* $L \leftrightarrow W \underset{H}{=} 1$.

Proof: Assume $W \underset{H}{=} 1$. Then by Lemma 18, $(1,W) \in L$ and so (s_3,s_3) is conjugate to $(s_3,W^{-1}s_3W)$ in L by $(1,W)$.

Conversely, assume (s_3,s_3) is conjugate to $(s_3,W^{-1}s_3W)$ in L. That is, for some $(X,Y) \in L$, $(X,Y)^{-1}(s_3,W^{-1}s_3W)(X,Y) \underset{G}{=} (s_3,s_3)$. From this equation it follows that

$$(X^{-1}s_3X,\ Y^{-1}W^{-1}s_3WY) \underset{G}{=} (s_3,s_3).$$

Hence, $X^{-1}s_3X \underset{F}{=} s_3$ and $(WY)^{-1}s_3(WY) \underset{F}{=} s_3$. But since F is free these equations imply (see [30]) that for some integers k,ℓ: $X \underset{F}{=} s_3^k$ and $WY \underset{F}{=} s_3^\ell$. Since $s_3 \underset{H}{=} 1$ it follows that $X \underset{H}{=} 1$ and $WY \underset{H}{=} 1$. By hypothesis $(X,Y) \in L$ and hence by Lemma 1 $X \underset{H}{=} Y$ so $Y \underset{H}{=} 1$ and $W \underset{H}{=} 1$ as claimed.||

Since H has unsolvable word problem, Lemma 24 completes the proof of Theorem 23.||

Let $GL(n,Z)$ denote the group of invertible $n \times n$ matrices with integer coefficients (such a matrix is invertible if and only if it has determinant ± 1). $SL(n,Z)$ denotes the normal subgroup of index 2 in

GL(n,Z) consisting of those $n \times n$ matrices with determinant +1. Note that SL(n,Z) is just the kernel of the determinant map from GL(n,Z) onto the multiplicative group of units in Z. Since the determinant of a matrix is effectively computable, it follows that SL(n,Z) is recursively embedded in GL(n,Z), i.e. has solvable GWP in GL(n,Z).

LEMMA 25: *Let* F *be a finitely generated free group. Then* F *can be recursively embedded in* SL(2,Z) *and* $F \times F$ *can be recursively embedded in* SL(n,Z) for $n \geq 4$.

Proof: We first show F can be recursively embedded in SL(2,Z). In view of Lemmas 19 and 20, it suffices to show the result when F has rank 2. Let T be the subgroup of SL(2,Z) generated by the matrices

$$\begin{pmatrix} 1 & 2 \\ 0 & 1 \end{pmatrix}, \quad \begin{pmatrix} 1 & 0 \\ 2 & 1 \end{pmatrix}.$$

Sanov [55] has shown that T is freely generated by these two matrices, and that an arbitrary 3×2 matrix

$$M = \begin{pmatrix} a & b \\ c & d \end{pmatrix}$$

belongs to T if and only if the following conditions are satisfied:

(1) $ad - bc = 1$

(2) a and d are congruent to 1 mod 4

(3) c and b are even.

Since one can effectively check these conditions, it follows that the map ϕ: $F \to T$ recursively embeds F in SL(2,Z).

Now $T \times T$ is recursively embedded in SL(4,Z) by

$$(U,V) \to \begin{pmatrix} U & O \\ O & V \end{pmatrix}$$

where $U,V \in T$. Moreover, $SL(4,Z)$ is recursively embedded in $SL(n,Z)$ for $n \geq 4$ by

$$X \to \begin{pmatrix} X & 0 \\ 0 & I \end{pmatrix}$$

where I is the $(n-4) \times (n-4)$ identity matrix and $X \in SL(4,Z)$. Hence, by Lemma 19, $F \times F$ can be recursively embedded in $SL(n,Z)$ for $n \geq 4$. This completes the proof.||

Applying Lemma 19, Theorem 21, and Theorem 23 we have the following two results:

THEOREM 26: *For* $n \geq 4$, $SL(n,Z)$ *has a finitely generated subgroup* L *such that: (1)* L *has unsolvable conjugacy problem and (2) the generalized word problem for* L *in* $SL(n,Z)$ *is unsolvable. Moreover, for each recursively enumerable degree* D, $SL(n,Z)$ *has a finitely generated subgroup whose generalized word problem in* $SL(n,Z)$ *has degree* D *of unsolvability.*||

COROLLARY 27: *Same as Theorem 26 with* $GL(n,Z)$ *in place of* $SL(n,Z)$.||

REMARKS: $GL(n,Z)$ has solvable word problem (just multiply out the matrices) and is residually finite. Finite presentations for $GL(n,Z)$ are known (see [30]). Recall also that $GL(n,Z)$ is just the automorphism group of the free abelian group of rank n.

Finally, combining Theorem 22 with Lemma 25 we obtain:

THEOREM 28: *The problem to determine of two finite sets of matrices whether or not they generate the same subgroup of* $SL(n,Z)(n \geq 4)$ *is recursively unsolvable.*||

D. ON A PROBLEM OF GRAHAM HIGMAN

Higman [25] has shown the existence of a universal finitely presented group U, i.e., a f.p. group U which contains an isomorphic copy of every f.p. group. In particular, U contains an isomorphic copy of every f.p. group with solvable word problem (WP), but U itself has unsolvable WP of degree $\underline{0}'$. Higman has asked the following question: Does there exist a universal f.p. "solvable WP" group, i.e., a f.p. group S with solvable WP which contains an isomorphic copy of every f.p. group with solvable WP? We show such a group does not exist.

The proof proceeds by contradiction. For suppose such a group S did exist. Let A(S) be the algorithm which solves the WP for S. Now let G be any f.p. group, say $G = \langle x_1,\ldots,x_n;\ R_1,\ldots,R_m \rangle$. Let $\phi_1, \phi_2, \ldots$ be an effective enumeration of all set maps from $\{x_1,\ldots,x_n\}$ into words of S. Extend each ϕ_i to products and inverses in a purely formal way. Now ϕ_i can be extended to a homomorphism of G into S if and only if $\phi_i(R_j) = 1$ in S for $1 \leq j \leq m$. But this can be recursively checked using A(S). Hence one can effectively enumerate all homomorphisms $\Psi_1, \Psi_2, \ldots$ from G into A(S).

Since G is f.p., the set of words $W = 1$ in G is effectively enumerable. Now suppose G has solvable WP. Then some Ψ_j is an embedding of G into S. Consequently, if $W \neq 1$ in G, then for some i, $\Psi_i(W) \neq 1$ in S, which can be tested using A(S). Hence the set of words $W \neq 1$ in G is effectively enumerable. This solves the word problem for G. (Note that we do not need to know which Ψ_j is an embedding).

Thus a partial algorithm has been given which, when applied to a f.p. group with solvable WP, solves the WP for that group. This contradicts a theorem of Boone and Rogers [15], and so the desired S could not exist.

REMARKS: Using a result of Murskij [36] the existence of a universal f.p. semi-group can easily be established. However, the above is easily modified to show the nonexistence of a universal "solvable WP" semi-group. In the semi-group case, defining relations are equations of the form $B_j = C_j$ and the word problem is to decide for an arbitrary pair of words W_1, W_2 whether or not $W_1 = W_2$. Homomorphisms are set maps which preserve products and defining equations. For fixed W_2, the set of words

W such that $W = W_2$ in the semi-group is r.e. The Boone-Rogers result holds for f.p. semi-groups. Hence the above argument will also work for semi-groups.

Finally, we remark that our method is closely related to the solution of the WP for residually finite groups given by [21] and [35].

CHAPTER IV

THE WORD AND CONJUGACY PROBLEMS FOR CERTAIN ELEMENTARY GROUPS

A. A GROUP THEORETIC CHARACTERIZATION OF TURING REDUCIBILITY

The following result gives a group-theoretic "characterization" of Turing reducibility for recursively enumerable degrees in the sense explained below:

THEOREM 1: *Let* S_1 *and* S_2 *be recursively enumerable sets of natural numbers. A necessary and sufficient condition that* S_1 *be Turing reducible to* S_2 *is that there exists a f.g. recursively presented group* G *whose word problem has the Turing degree of* S_1 *and whose conjugacy problem has the Turing degree of* S_2.

REMARKS: Actually, we derive a slightly stronger result than Theorem 1 which will be formulated later. Let D_1 be the degree of S_1 and D_2 be the degree of S_2. Assume that $D_1 \leq D_2$. The construction is given in two stages: first, a group T is constructed which has solvable word problem, but conjugacy problem of degree D_2; then a similar group R is constructed which has word and conjugacy problem both of degree D_1. Finally, we simply define $G = T * R$ and the result follows.

Theorem 1 gives a group-theoretic "characterization" of Turing reducibility for r.e. Turing degrees in the following sense: Assume the r.e. Turing degrees are given *a priori* qua classes of sets of integers. Then for D_1, D_2 r.e. degrees, $D_1 \leq_T D_2$ if and only if, for some f.g. recursively presented group G, the word problem for G belongs to D_1 and the conjugacy problem for G belongs to D_2.

Since the word problem is always one-one reducible to the conjugacy problem, it is quite possible that an analagous theorem could hold for some other kind of reducibility and degrees, e.g. truth table or many-one (not one-one, however, since the one-one degree of a simple set does not contain a word problem). In fact, for a suitable interpretation of our proof, the word problem for G of Theorem 1 has the same truth-table degree as S_1. However, taking $S_1 \leq_T S_2$ but S_1 not truth table reducible to S_2, it follows that the conjugacy problem for G can not have the same truth table as S_2. For otherwise, by transitivity of truth table reducibility, S_1 would be truth table reducible to S_2. In view of our remarks about the method of construction above, we actually show S_1 join S_2 has the same Turing degree as the conjugacy problem for G. In any event, the analogue of Theorem 1 might hold for some other kind of reducibility, but some other method of proof would be required.

Another natural conjecture is that G can be chosen finitely presented.[1]

Finally we mention that the proofs in this chapter (sections A and B) are not "constructive" in the sense that we must use an enumeration to find numbers or words known to exist by having asked an oracle. Thus, knowing something exists using an oracle, we do not yet have the object but must find it by enumerating some r.e. set (See Boone [11] in this connection).

Proof of Theorem 1: Let D_1 be the degree of S_1 and D_2 the degree of S_2. The "sufficiency" assertion of Theorem 1 is obvious. Assume $D_1 \leq D_2$. Let N denote the natural numbers. Let g be a one-one total

[1] Added in proof: D. J. Collins has recently established this conjecture (see [58]).

recursive function with infinite range $g(N) \epsilon D_2$ and $O \not\epsilon g(N)$. (Clearly any r.e. degree can be represented by such a function).

Define a new function f as follows:

$$f(2i) = 2g(i)$$
$$f(2i+1) = 2g(i) + 1.$$

Then f is a one-one total recursive function with infinite range, $O \not\epsilon f(N)$ and $f(N) \epsilon D_2$. Let $B = < c,d,e >$ be a free group and H its infinitely generated subgroup

$$H = < e^{-i}c^{-f(i)}d\ c^{f(i)}e^{i},\ i > O >.$$

Let $\overline{B} = < \overline{c},\overline{d},\overline{e} >$ be a distinct copy of B with $\overline{H}$ the copy of H in $\overline{B}$. Now form $T = (B * \overline{B};\ H = \overline{H})$. Since f is a recursive function, T is a recursively presented group.

Observe that the listed generators for H are indeed a Nielsen basis for H, e.g., in the product $(e^{-i}c^{-f(i)}d\ c^{f(i)}e^{i})(e^{-j}c^{-f(j)}d\ c^{f(j)}e^{j})$ for $i \neq j$ there is no cancellation except among the interior e-segments. Since f is recursive, it is clear that one can inductively decide for an arbitrary word $W \epsilon B$ whether or not $W \epsilon H$. The amalgamating isomorphism is recursively defined (just the "bar" map), so it follows from [30] Lemma 4.9, p. 272, that the word problem for T is solvable.

The notation $W \sim H$ for a word W means that W is conjugate to an element of the subgroup H. For the present we investigate the group B and its subgroup H.

LEMMA 2: *The problem of deciding for arbitrary* $W \epsilon B$ *whether or not* $W \sim H$ *is equivalent to the membership problem for* f(N), *and hence has degree* D_2.

Proof: We first show the problem is reducible to the membership problem for f(N). An arbitrary cyclically reduced element $V \epsilon H$ (cyclically reduced in H) has form

$$(1)\ V = e^{-\alpha_1}c^{-f(\alpha_1)}d^{\beta_1}c^{f(\alpha_1)}e^{\alpha_1-\alpha_2} \cdots e^{\alpha_{n-1}-\alpha_n}c^{-f(\alpha_n)} d^{\beta_n}c^{f(\alpha_n)}e^{\alpha_n}.$$

where $\beta_i \neq 0$, $\alpha_i \neq \alpha_{i+1}$ and $\alpha_1 \neq \alpha_n$ (since f is one-one). This word is conjugate in B for $n > 1$ to the cyclically reduced form (in B).

$$(2)\ e^{\alpha_1}Ve^{-\alpha_1} = c^{-f(\alpha_1)}d^{\beta_1}c^{f(\alpha_1)}e^{\alpha_1-\alpha_2} \cdots e^{\alpha_{n-1}-\alpha_n}c^{-f(\alpha_n)} d^{\beta_n}c^{f(\alpha_n)}e^{\alpha_n-\alpha_1}.$$

Observe that the exponent sum on V or its conjugates must be 0 for e and c and $\beta_1 + \ldots + \beta_n$ for d.

By [30] Theorem 1.3, p. 36, an arbitrary word $W \in B$ has $W \sim H$ only if on applying some cyclic permutation π to the cyclic reduction $\sigma(W)$ of W one has

$$(3)\ \pi\ \sigma(W) = c^{-\gamma_1}d^{\beta_1}c^{\gamma_1}e^{\delta_1}c^{-\gamma_2}d^{\beta_2} \cdots e^{\delta_{n-1}}c^{-\gamma_n}d^{\beta_n}c^{\gamma_n}e^{\delta_n}.$$

where $\delta_1 + \ldots + \delta_n = 0$ as the exponent sum on e must be 0. (In the exceptional case $\pi\ \sigma(W) = d^{\beta_1}$ corresponding to $n = 1$ in (1) above, the answer is clearly that $W \sim H$. Hence only $n > 1$ need be considered hereafter so (3) applies).

Call a word of the form (3) an *allowable permutation* of $\sigma(W)$. Note one can constructively find all allowable permutations of $\sigma(W)$ (there are only finitely many permutations of a word). If there are no allowable permutations of $\sigma(W)$, then $W \nsim H$. Note that the conjugation of some allowed permutation $\pi\ \sigma(W)$ into H can be done by e^i for some i if it can be done at all.

Suppose $\pi\ \sigma(W)$ is allowable. Decide for $\gamma_1, \ldots, \gamma_n$ successively whether $\gamma_i \in f(N)$. If any $\gamma_i \notin f(N)$, then clearly $\pi\ \sigma(W)$ and hence W is not conjugate to an element of H. On the other hand, suppose all $\gamma_i \in f(N)$. Then by enumerating f(N) we can find the unique $\alpha_i \in N$

such that $f(\alpha_i) = \gamma_i$, $i = 1,\ldots, n$. In view of (1), $e^{-\alpha_1}\pi\ \sigma(W)e^{\alpha_1} \in H$ if and only if

$$\delta_1 = \alpha_1 - \alpha_2, \ldots, \delta_{n-1} = \alpha_{n-1} - \alpha_n, \delta_n = \alpha_n - \alpha_1;$$

this can be effectively tested.

Finally, one allowable permutation is conjugate to an element of H by an e^i if and only if every allowable permutation is. Hence the above procedure reduces deciding whether $W \sim H$ to membership for f(N).

Next we show that a procedure to decide whether $W \sim H$ would yield a procedure to decide membership in f(N). Suppose we have a procedure to decide whether $W \sim H$. Now let

$$V_m = c^{-2m}d\, c^{2m}e^{-1}c^{-2m-1}d\, c^{2m+1}e.$$

(where $m > 0$). Now as above $V_m \sim H$ if and only if $\exists\ i > 0$ such that $e^{-i}\, V_m\, e^i \in H$. By properties of f and H, it follows that $V_m \sim H$ if and only if $m \in g(N)$. Hence a procedure to decide whether $V_m \sim H$ would yield a procedure to decide whether $m \in g(N)$ which is equivalent to deciding membership in f(N).||

COROLLARY 3: *The conjugacy problem for* T *has degree* $\geq D_2$.

Proof: Let $W \in B$ and $\overline{W} \in \overline{B}$, the corresponding word with bars. Then by Solitar's Theorem, $W \sim \overline{W}$ in T if and only if $W \sim H$. For then $\overline{W} \sim \overline{H} = H$ and W and $\overline{W}$ are conjugate to the same element of $H = \overline{H}$. Hence a decision procedure for the conjugacy problem would yield a decision procedure for deciding whether $W \sim H$. Lemma 2 now gives the result.||

Rewrite the relations for T in the following form: $d\, c^{f(i)}e^i\overline{e}^{-i}\overline{c}^{-f(i)}\overline{d}^{-1} = c^{f(i)}e^i\overline{e}^{-i}\overline{c}^{-f(i)}$, $(i > 0)$. Put $K = \langle c, e, \overline{c}, \overline{e} \rangle$, a free group. From the above relations it is immediate that:

LEMMA 4: T *is a Britton extension of* K *with respect to the stable letters* $\{d, \overline{d}\}$.||

Moreover, because of these relations it follows that:

LEMMA 5: *Two* $\{d,\bar{d}\}$ – *free words of* T *are conjugate* $\leftrightarrow$ *they are conjugate in* K. *Hence, whether or not two* $\{d,\bar{d}\}$ – *free words are conjugate is decidable.*||

By the *length* of a word $W \in T$ we mean the normal form length in T qua element of a free product with amalgamation. Since we can solve the word problem for T and the generalized word problem for H in B from [30] exercise 26, p. 104, a minimal Schreir system of coset representatives for H in B can be found and the normal form of a word W can be effectively computed. Hence, there is an effective procedure which cyclically reduces any $W \in T$ in the sense of amalgamated free products.

LEMMA 6: *Let U be a cyclically reduced word of* T *which has length* < 2. *There is a procedure (uniform in* U*) which reduces the problem of deciding for arbitrary* $V \in T$ *whether or not* $U \underset{T}{\sim} V$ *to the membership problem for* f(N).

Proof: First (effectively) replace V by its cyclic reduction V_0. If length $V_0 \geq 2$, then $V_0 \not\sim U$. So assume length $V_0 < 2$. Now U lies in some factor, say B. Assume hereafter that we have an oracle for deciding membership in f(N). Decide (via Lemma 2) whether $U \sim H$. *Case 1:* $U \not\sim H$. Decide whether $V_0 \in B$. If not, then $V_0 \not\sim U$ by Solitar's Theorem. Assume $V_0 \in B$. Then by Solitar's Theorem, $U \sim V_0 \leftrightarrow U \underset{B}{\sim} V_0$ which is decidable. *Case 2:* $U \sim H$. Decide whether $V_0 \sim H$. If not, then $V_0 \not\sim U$ by Solitar's Theorem. Assume $V_0 \sim H$. Then replace U, V_0 by their conjugates U', $V_0' \in H$. Since the actions of B and $\bar{B}$ on elements of $H = \bar{H}$ are the same, it follows that $U' \underset{T}{\sim} V_0' \leftrightarrow U' \underset{B}{\sim} V_0'$ which is decidable.||

Recall that T is a Britton extension of K (Lemma 4). The words $c^{f(i)} e^i \bar{e}^{-i} \bar{c}^{-f(i)}$ $(i > 0)$ are a Nielsen basis for the subgroup of K they generate. Consequently, one can decide whether an element of K lies in that subgroup and hence can effectively $\{d,\bar{d}\}$ – reduce any word of T. Moreover, it is easy to see that $\{d,\bar{d}\}$ – reductions do not change the normal form length of a word of T. Thus, for any word $U \in T$ we can effectively find a cyclically reduced and $\{d,\bar{d}\}$ – cyclically reduced word of T which is conjugate to U.

LEMMA 7: *Let* U *be a cyclically reduced word of* T *which has length* ≥ 2. *There is an effective procedure (uniform in* U*) which decides for arbitrary* $V \in T$ *whether or not* $U \underset{T}{\sim} V$.

Proof: By using the previously described procedures we may assume that both U and V are cyclically reduced and $\{d,\bar{d}\}$ – cyclically reduced. If both are $\{d,\bar{d}\}$ – free, then conjugacy is decided by Lemma 5. If only one is $\{d,\bar{d}\}$ – free, then $U \not\sim V$ by Collins' Lemma. Hence we may assume that both U and V involve d or $\bar{d}$.

By Collins' Lemma, if $U \sim V$ there is a cyclic permutation V_0 of V such that U and V_0 are $\{d,\bar{d}\}$ – parallel. Determine by inspection if the $\{d,\bar{d}\}$ – structure of V makes this possible. If not, $V \not\sim U$. Thus assume this is possible.

By hypothesis U and V are cyclically reduced in T. Assume $U \sim V$. Then there is a cyclic permutation V_0 of V such that $U = W^{-1}V_0W$ where W lies in the subgroup H, by Solitar's Theorem. Denote the i-th listed generator $e^{-i}c^{-f(i)}d\, c^{f(i)}e^{i}$ of H by Y_i and the word $c^{f(i)}e^{i}\bar{e}^{-i}\bar{c}^{-f(i)}$ by X_i. Write

$$U \equiv \Psi_1 \tilde{d}^{\varepsilon_1} \Psi_2 \ldots \Psi_n \tilde{d}^{\varepsilon_n} \Psi_{n+1}$$

$$V_0 \equiv \theta_1 \tilde{d}^{\delta_1} \theta_2 \ldots \theta_n \tilde{d}^{\delta_n} \theta_{n+1}$$

where $\tilde{d}$ is a variable for d or $\bar{d}$, ε_i, $\delta_i = \pm 1$ and Ψ_i, θ_i are $\{d;\bar{d}\}$ – free.

Before continuing the proof, consider the following example calculation:

$$\text{Let } V_0 = Y_5^{-1}Y_5^{-1}e^{-5}c^{-f(5)}X_3\,\overline{d}^{-1}X_1^{-1}c^{f(2)}e^2.$$

Then $Y_2^{-1}V_0Y_2 = Y_2^{-1}Y_5^{-1}Y_5^{-1}e^{-5}c^{-f(5)}X_3\,\overline{d}^{-1}X_1^{-1}c^{f(2)}e^2e^{-2}c^{-f(2)}d\,c^{f(2)}$

$$Y_2^{-1}Y_5^{-1}Y_5^{-1}e^{-5}c^{-f(5)}X_3X_1^{-1}c^{f(2)}e^2.$$

Now $Y_2^{-2}V_0Y_2{}^2 = Y_2^{-2}Y_5^{-1}Y_5^{-1}e^{-5}c^{-f(5)}X_3X_1^{-1}c^{f(2)}e^2e^{-2}c^{-f(2)}d\,c^{f(2)}e^2$

$Y_2^{-2}Y_5^{-1}e^{-5}c^{-f(5)}d^{-1}X_3X_1^{-1}d\,c^{f(2)}e^2 = Y_2^{-2}Y_5^{-1}e^{-5}c^{-f(5)}X_3X_1^{-1}c^{f(2)}e^2.$

Similarly,

$$Y_2^{-3}\,V_0\,Y_2^3 = Y_2^{-3}\,e^{-5}c^{-f(5)}\,X_3X_1^{-1}c^{f(2)}e^2.$$

This example may help in understanding what follows.

Now W is a word on the Y_i and we may assume W is freely reduced, hence $\{d,\overline{d}\}$ – reduced. By assumption, $U = W^{-1}V_0W$. Let Y_j^{ε} be the first Y_i of W. Then, in $Y_j^{-\varepsilon}V_0\,Y_j^{\varepsilon}$ either the d^{ε} in Y_j or the $d^{-\varepsilon}$ in $Y_j^{-\varepsilon}$ (not both) must pinch against $\tilde{d}^{\delta_n}$ or $\tilde{d}^{\delta_1}$ respectively.

Case 1: Assume d^{ε} pinches against $\tilde{d}^{\delta_n}$. Then, since $Y_j^{\varepsilon} = e^{-j}\,c^{-f(j)}\,d^{\varepsilon}\,c^{f(j)}e^j$ we must have $\theta_{n+1}e^{-j}c^{-f(j)}$ = word on X_i = $c^{f(i)}e^i\overline{e}^{-i}\overline{c}^{-f(i)}$. Consequently $\theta_{n+1} \equiv \theta\,c^{f(j)}\,e^j$ without cancellation where θ is a word on the X_i. Hence Y_j^{ε} can be effectively determined from θ_{n+1} and $\tilde{d}^{\varepsilon_n}$; also $Y_j^{-\varepsilon}\,V_0\,Y_j^{\varepsilon} = Y_j^{-\varepsilon}\theta_1\tilde{d}^{\delta_1}\,\theta_2\,\ldots\,\tilde{d}^{\delta_{n-1}}\theta_n\theta_{n+1}$. But, if Y_m^{γ} is the next Y_i in W, since the $d^{-\gamma}$ in $Y_m^{-\gamma}$ cannot pinch against the $d^{-\varepsilon}$ in $Y_j^{-\varepsilon}$, again it follows that $Y_m^{-\gamma}$ is determined effectivel from $\tilde{d}^{\delta_{n-1}}$ and $\theta_n \theta_{n+1}$. Continuing in this way, it follows that first n of the $Y_i^{\pm 1}$ in W can be effectively found from the given word V_0. (assuming W contains this many Y_i's).

Assume W contains more than n of the Y_i's. Write $W = W_0W_1$ where W_0 is the product of the first n Y_i's in W. Now

$$W_0^{-1} V_0 W_0 = W_0^{-1} \theta_1\theta_2 \dots \theta_{n+1}.$$

Since W_1 is non-empty, the first Y_ℓ^ε in W_1 must have d^ε pinching against the $d^{-\varepsilon}$ appearing last in $Y_j^{-\varepsilon}$ in W_0^{-1}. Consequently,

$$Y_\ell^{-\varepsilon} W_0^{-1} V_0 W_0 Y_\ell^\varepsilon = Y_\ell^{-\varepsilon} W_0^{-1} \theta_1\theta_2 \dots \theta_{n+1} Y_\ell^\varepsilon =$$

$$Y_\ell^{-\varepsilon} W_0^{-1} Y_j^\varepsilon \theta_1\theta_2 \dots \theta_{n+1}.$$

But again Y_ℓ^ε can be effectively determined from $c^{f(j)} e^j \theta_1\theta_2 \dots \theta_{n+1}$. Indeed, $c^{f(j)} e^j \theta_1\theta_2 \dots \theta_{n+1} e^{-\ell} c^{-f(\ell)}$ = word on $X_i = c^{f(i)} e^i \bar{e}^{-i} \bar{c}^{-f(i)}$. Let $\theta_0 = \theta_1\theta_2 \dots \theta_{n+1}$ be freely reduced. Then $\theta_0 = e^{-j} c^{-f(j)} \Gamma c^{f(\ell)} e^\ell$ where Γ is a (non-empty) word on the X_i (not necessarily without cancellation). So $Y_\ell^{-\varepsilon} W_0^{-1} V_0 W_0 Y_\ell^\varepsilon = Y_\ell^{-\varepsilon} W_0^{-1} Y_j^\varepsilon \theta_0$.

In view of the form of θ_0 we see that, W_1 must be a power of Y_ℓ^ε in order for more pinchings to occur. Indeed the highest power is at most n provided W_1 is chosen of minimal power, for after that further conjugation by Y_ℓ would repeat a previous equation.

From all of this we see that $U = W^{-1} V_0 W$ implies there exists such a W of length at most 2n in the Y_i where the Y_i in W can be effectively determined from V_0.

Case 2: Assume d^{ε} pinches against $\tilde{d}^{\delta_1}$. The dual argument (right and left interchanged) leads to the same conclusion: $U = W^{-1}V_0W$ implies there exists such a W of length at most $2n$ in the Y_i where the Y_i in W can be effectively determined from V_0. We omit the dual proof.

Finally, since there are only finitely many candidates for V_0, there is an effective procedure to decide for arbitrary $V \in T$ whether or not $U \underset{T}{\sim} V$ (uniformly in U).||

As a result of the above lemmas and discussion we conclude:

THEOREM 8: *The f.g. recursively presented group* T *has solvable word problem but conjugacy problem of degree* D_2.||

Now let M be a set of integers, $O \notin M \in D_1$. Put

$$L = < e^{-m}c^{-m}d\ c^m e^m,\ m \in M >,$$

a free subgroup of B. Let $\bar{L}$ be the corresponding subgroup under the "bar" map ψ. Finally, let $R = (B * \bar{B}; L = \bar{L}, \Psi)$. From the argument given by Boone in [10] it follows that:

LEMMA 9: *The word problem for* R *has degree* D_1. *Hence the conjugacy problem for* R *had degree* $\geq D_1$.||

All of the arguments above for T remain valid for R provided we replace "solvable" by "D_1-solvable", "effective" by "D_1-effective", "constructive" by "D_1-constructive", etc. (The part of Lemma 2 depending on the special functions f and g is no longer necessary in view of Lemma 9). We will not repeat the arguments, but simply draw the desired conclusions:

LEMMA 10: *The conjugacy problem for* R *has degree* D_1. *So* R *has word and conjugacy problems both of degree* D_1.||

Observe that there is no necessity for D_1 to be an r.e. degree. Putting $G = T * R$ (or alternatively $G = T \times R$), we have proved the following result which implies Theorem 1:

THEOREM 11: *Let* D_1, D_2 *be degrees with* D_2 *r.e. Then there exists a finitely generated group* G *with word problem of degree* D_1 *and conjugacy problem of degree* = *lub* (D_1, D_2). *If* D_1 *is r.e.,* G *can be recursively presented.*

B. DEGREE RESULTS FOR THE CONJUGACY PROBLEM IN ELEMENTARY GROUPS

In Chapter III an example of a f.p., residually finite, strong Britton extension of a free group with unsolvable conjugacy problem was given. In this section the following arbitrary r.e. degree analogue of that result is obtained.

THEOREM 12: *Let* $H = < s_1, \ldots, s_n; R_1, \ldots, R_m >$ *be a f.p. group with word problem of degree* D. *There is an explicit uniform method for constructing from* H *the following: (1) a f.p., residually finite, Britton extension* G *of a free group with solvable word problem, but conjugacy problem of degree* D; *and (2) the free product with f.g. amalgamation of two free groups which has solvable word problem, but conjugacy problem of degree* D.

In view of the discussion in Chapter II - Section C, the second assertion (2) follows immediately from the first (1). The remainder of this section is devoted to establishing (1).

Let $F = < s_1, \ldots, s_n, \bar{s}_1, \ldots, \bar{s}_n, q >$, a free group on $2n + 1$ generators.

The symbol $\tilde{s}$ will be a variable for an s_α or an $\bar{s}_\alpha$.

Define G to be the f.p. group with:

Generators: $s_1, \ldots, s_n, \bar{s}_1, \ldots, \bar{s}_n, q, t_1, \ldots, t_m$

$a_1, \ldots, a_n, b_1, \ldots, b_n, c_1, \ldots, c_n.$

$$\text{Relations: } \left.\begin{aligned} t_i^{-1}\, q\, t_i &= q\, R_i \\ t_i^{-1}\, \tilde{s}\, t_i &= \tilde{s} \end{aligned}\right\} \quad \begin{aligned} &1 \le i \le m \\ &\text{each } \tilde{s} \end{aligned}$$

$$\left.\begin{aligned} a_\alpha^{-1}\, \bar{s}_\alpha\, q\, a_\alpha &= q\, s_\alpha \\ a_\alpha^{-1}\, \tilde{s}\, a_\alpha &= \tilde{s} \end{aligned}\right\} \quad \begin{aligned} &1 \le \alpha \le n \\ &\text{each } \tilde{s} \end{aligned}$$

$$\left.\begin{aligned} b_\alpha^{-1}\, \tilde{s}\, b_\alpha &= \tilde{s} \\ b_\alpha^{-1}\, q\, c_\alpha &= q\, s_\alpha \\ c_\alpha^{-1}\, q\, b_\alpha &= q\, s_\alpha \\ c_\alpha^{-1}\, \tilde{s}\, c_\alpha &= \tilde{s} \end{aligned}\right\} \quad \begin{aligned} &1 \le \alpha \le n \\ &\text{each } \tilde{s}. \end{aligned}$$

This G is the desired group. Note that the example given in III-A is defined in a similar, but less complicated manner.

Denote by P the collection of letters

$$\{t_1,\ldots,t_m,\ a_1,\ldots,a_n,\ b_1,\ldots,b_n,\ c_1,\ldots,c_n\}.$$

LEMMA 13: G *is a Britton extension of the free group* F *with respect to the stable letters* P.

Proof: One easily checks that each of the subgroups (for equivalent P-symbols) in question is F and is generated by the $2n + 1$ elements listed in the relations – hence they freely generate F and the GIC is satisfied.||

By Lemma II-9 it follows that:

LEMMA 14: G *has solvable word problem and there is an effective procedure for P-reducing words of* G.||

The relations of G involving q are:

$$q^{-1}\, t_i\, q = t_i\, R_i^{-1} \qquad (1 \le i \le m)$$

$$\left.\begin{aligned} q^{-1}\, \bar{s}_\alpha^{-1}\, a_\alpha\, q &= a_\alpha s_\alpha^{-1} \\ q^{-1}\, b_\alpha\, q &= c_\alpha s_\alpha^{-1} \\ q^{-1}\, c_\alpha\, q &= b_\alpha s_\alpha^{-1} \end{aligned}\right\} \qquad 1 \le \alpha \le n$$

Let J be the free group on $s_1, \ldots, s_n, \bar{s}_1, \ldots, \bar{s}_n$, and let L be the free group on the letters of P. Put $K = J \times L$.

LEMMA 15: G *is a strong Britton extension of* K *with respect to* q. K *has solvable word problem. There is an effective method for q-reducing a word of* G.

Proof: By deleting q from the generators of G and deleting the relations involving q from the relations of G, one obtains a presentation of K. Clearly, the letters of P freely generate a free subgroup of K, and K has solvable word problem. Moreover, G is a strong Britton extension of K with respect to q since each of the two subgroups in question is free on the same number of generators. Finally to q-reduce a word of G, it is enough to decide whether an element of K lies in the correct subgroup. But these subgroups are free, and a word in the subgroup is completely determined by its P-structure. Hence, using the solution to the word problem for K, one can decide whether a word of K belongs to the correct subgroup.‖

NOTATION: Let X, Y, Z be variables for words in the s_α. If $X \equiv s_{\alpha_1}^{\varepsilon_1} \ldots s_{\alpha_k}^{\varepsilon_k}$ write $\bar{X}$ for the word $\bar{s}_{\alpha_1}^{\varepsilon_1} \ldots \bar{s}_{\alpha_k}^{\varepsilon_k}$ and write $\tilde{X}$ for X or $\bar{X}$ usually depending on the cases under consideration. The functional notation $W(t_i, \bar{s}_\alpha^{-1} a_\alpha, b_\alpha, c_\alpha)$ means that W is a word which is a product of t_i, $\bar{s}_\alpha^{-1} a_\alpha$, b_α, c_α or their inverses for $1 \le i \le m$, $1 \le \alpha \le n$.

$W(t_iR_i^{-1}, a_\alpha s_\alpha^{-1}, c_\alpha s_\alpha^{-1}, b_\alpha s_\alpha^{-1})$ will denote the corresponding word obtained by formal substitutions. Observe that these two words are conjugate by q.

Inspecting the defining relations for G, we observe the following three lemmas:

LEMMA 16: *Any word* $W \in G$ *can be effectively written as* $W = U\theta$ *where* $U \in F$ *and* $\theta \in L$, *i.e.,* $G = F \cdot L.$ ||

LEMMA 17: *If* θ *is a word on the* t_i *and* a_α, *then there are words* X, Y *such that* $\theta^{-1}\, q\, \theta = \overline{X}\, q\, Y$ *and* $\theta^{-1}\, \tilde{s}\, \theta = \tilde{s}$ *where* $X\,Y \underset{H}{=} 1.$ ||

LEMMA 18: *Let* $U \in F$. Then in the words $b_\alpha^{-\varepsilon}\, U\, b_\alpha^{\varepsilon}$ *and* $c_\alpha^{-\varepsilon} U\, c_\alpha^{\varepsilon}$, *the* b_α *and* c_α *can be pinched out to give a P-free word* $\leftrightarrow$ U *contains an even number of q-symbols (possibly zero).* ||

The following result shows the connection between conjugacy in G and equality in H.

LEMMA 19: (1) $\overline{X}_1\, q\, Y_1 \underset{G}{\sim} \overline{X}_2\, q\, Y_2 \leftrightarrow X_1Y_1 \underset{H}{=} X_2Y_2$. *Moreover*

(2) $\overline{X}_1 q Y_1 \underset{G}{\sim} \overline{X}_2 q Y_2 \leftrightarrow$ *there is a word* $W(t_i, a_\alpha)$ *on the* t_i *and* a_α *alone such that* $\overline{X}_1\, q\, Y_1 \underset{G}{=} W^{-1}\, \overline{X}_2\, q\, Y_2\, W.$

Proof: (2) $\leftarrow$ is trivial. Consider (1) $\leftarrow$: The assumption $X_1Y_1 \underset{H}{=} X_2Y_2$ implies that

$$X_1Y_1 = X_2\Big(\prod_{i=1}^{k} Z_i\, R_{\gamma_i}^{\varepsilon_i}\, Z_i^{-1}\Big)Y_2$$

in the free group $< s_1,\ldots,s_n >$ since $X_2^{-1}X_1Y_1Y_2^{-1} \underset{H}{=} 1.$

Let $Z_i(s_\alpha)$ be Z_i and let $\theta_i = Z_i(a_\alpha^{-1})t_{\gamma_i}^{\varepsilon_i}\ Z_i(a_\alpha^{-1})^{-1}$.

One easily calculates that

$$Z_i(a_\alpha^{-1})^{-1}\ q\ Z_i(a_\alpha^{-1}) = \bar{Z}_i\ q\ Z_i^{-1}.$$

And $$\theta_i^{-1}\ q\ \theta_i = q\ Z_i\ R_{\gamma_i}^{\varepsilon_i}\ Z_i^{-1}.$$

Put $$W_1 = \theta_k\theta_{k-1}\cdots\theta_2\theta_1.$$

Now one calculates that

$$W_1^{-1}\ q\ W_1 \underset{G}{=} q(\prod_{i=1}^{k} Z_i\ R_{\gamma_i}^{\varepsilon_i}\ Z_i^{-1}).$$

But the equations

$$X_1(a_\alpha^{-1})\bar{X}_1\ q\ Y_1X(a_\alpha^{-1})^{-1} \underset{G}{=} q\ X_1Y_1$$

$$X_2(a_\alpha^{-1})\ q\ X_2(a_\alpha^{-1})^{-1} \underset{G}{=} \bar{X}_2^{-1}\ q\ X_2$$

are easily verified. So letting $W = W_1X_2(a_\alpha^{-1})^{-1}\ X_1(a_\alpha^{-1})$ and using the fact that t_i and a_α commute with $\tilde{s}$ symbols, it follows by direct calculation that $\bar{X}_1\ q\ Y_1 = W^{-1}\bar{X}_2\ q\ Y_2\ W$ as claimed. Notice that W is a word on the t_i and a_α alone.

Next consider (1) → : Assume $\bar{X}_1qY_1 \underset{G}{\sim} \bar{X}_2qY_2$. Both words are q-reduced so by Collins' Lemma $q\ Y_1\bar{X}_1 = W^{-1}\ q\ Y_2\bar{X}_2\ W$ where W is a word on t_i, $\bar{s}_\alpha^{-1}\ a_\alpha$, b_α and c_α. But by Theorem A and Lemma 18, W is a word on the t_i and $\bar{s}_\alpha^{-1}a_\alpha$ alone. Now $W \underset{K}{=} \bar{Z}\ \theta$ where θ is a word on the t_i and a_α. In view of Lemma 17, $\theta^{-1}q\ \theta = \bar{X}_0\ q\ Y_0$ where

$X_0Y_0 \underset{H}{=} 1$. Now we have $W^{-1}\ q\ Y_2\bar{X}_2W \underset{G}{=} \bar{Z}^{-1}\bar{X}_0\ q\ Y_0Y_2\bar{X}_2\bar{Z} \underset{F}{=} q\ Y_1\bar{X}_1$.

Since the last equality holds in the free group F we conclude: $Z = X_0$, $X_1 = X_2Z = X_2X_0$, and $Y_1 = Y_0Y_2$ in the free group on the s_α. But $X_0Y_0 \underset{H}{=} 1$, so $X_1Y_1 \underset{F}{=} X_2X_0Y_0Y_2 \underset{H}{=} X_2Y_2$ as claimed.

Finally (2) $\rightarrow$ follows from (1) $\rightarrow$ and the proof of (1) $\leftarrow$. ||

Since H has word problem of degree D, by Lemmas 14 and 19 it follows that:

LEMMA 20: G *has solvable word problem, but conjugacy problem of degree* $\geq$ D. ||

The following result gives essential information about the b_α and c_α:

LEMMA 21: *Let* X, Y *be arbitrary words on the* s_α. *Then one can effectively find a word* $W(a_\alpha, b_\alpha, c_\alpha)$ *such that*

$$W^{-1}(a_\alpha,b_\alpha,c_\alpha)q\ W(a_\alpha,c_\alpha,b_\alpha) = \bar{X}\ q\ Y.$$

(note this is not a conjugacy equation since c_α *and* b_α *are interchanged). Moreover, if* $V(q,\tilde{s})$ *is any word of* F *with an even number of q-symbols, then*

$$W^{-1}\ V(q,\tilde{s})W = V(\bar{X}\ q\ Y,\ \tilde{s})$$

where W *is* $W(a_\alpha,b_\alpha,c_\alpha)$. *(This is a conjugacy equation).*

Proof: In view of the defining relations for G and Lemma 18, the second claim follows directly from the first claim. To prove the first claim, let $Z(s_\alpha) = XY$. Now one calculates that

$$Z(b_\alpha^{-1})\ q\ Z(c_\alpha^{-1})^{-1} \underset{G}{=} q\ Z \underset{F}{=} q\ XY.$$

Now $$X(a_\alpha^{-1})^{-1}qXY\ X(a_\alpha^{-1}) = \bar{X}\ q\ X^{-1}XY = \bar{X}\ q\ Y,$$

so putting $W(a_\alpha,b_\alpha,c_\alpha) = Z(b_\alpha^{-1})^{-1}\ X(a_\alpha^{-1})$ the result follows by an obvious calculation. ||

Let W be a word of G. Observe that P-pinches on W preserve the number of q-symbols in W and q-pinches preserve the number of P-symbols in W. Consequently, we may define a word to be (P,q)-reduced if it is both P-reduced and q-reduced. Similarly for (P,q)-cyclically reduced. Moreover, there is an effective process which computes, for any $W \in G$, a (P,q)-cyclically reduced word W_0 which is conjugate to W in G. (For more formal details see Collins [18]).

The proof that G is residually finite will be deferred until we establish:

THEOREM 22: *The conjugacy problem for* G *is reducible to the word problem for* H *and hence has degree* D.

Proof: Assume that an oracle (degree D) to solve the word problem for H is given. Uses of the oracle will be marked by (#). Let U, V be two given words of G. We will eventually show how to decide whether or not $U \underset{G}{\sim} V$. Without loss of generality, U,V are assumed to be (P,q)-cyclically reduced. The conventions regarding the use of Collins' Lemma explained in II-B will be used frequently in what follows.

LEMMA 23: *Let* U, V *be* (P,q)-*free words of* G. *Then* $U \underset{G}{\sim} V$ *if and only if* U *and* V *are conjugate in* J *(i.e. freely conjugate).*

Proof: The sufficiency is trivial. For the necessity, assume $\exists\, W \in G$ such that $U = W^{-1}\, V\, W$. If W is (P,q)-free, the conclusion is immediate. The proof proceeds by induction on the number of (P,q)-symbols in W. We may assume W is (P,q)-reduced. It will be shown that W can be replaced by a conjugating word W_0 with fewer (P,q)-symbols, thus completing the induction.

Let y^{ε} be the first (P,q)-symbol of W. Then $W \equiv \theta\, y^{\varepsilon}\, W_1$ where θ is a word on the $\tilde{s}$. Examine the equation $U = W_1^{-1}\, y^{-\varepsilon}\, \theta^{-1}\, V\, \theta y^{\varepsilon}\, W_1$; it is clear that $\theta^{-1} V\, \theta$ is a word on the $\tilde{s}$. By Theorem A, y^{ε} must pinch out across $\theta^{-1}\, V\, \theta$. In case y^{ε} is q^{ε}, this implies $\theta^{-1}\, V\, \theta \underset{J}{=} 1$ and so

$U \underset{J}{=} 1$ and the lemma follows since $U \underset{J}{=} V$. In case y^{ε} is one of t_i, a_α, b_α, c_α, from the relations for G, y^{ε} commutes with $\theta^{-1} V \theta$; hence U is conjugate to V by θW_1, a word with fewer (P,q)-symbols.‖

Case 1: In view of Lemma 23 and Collins' Lemma, if U is (P,q)-free, whether or not $U \underset{G}{\sim} V$ is a solvable problem. For then V must be (P,q)-free, and J has a solvable conjugacy problem.

Case 2 (#): Next consider the case that U is P-free, but involves a single q, which we may assume is q^{+1} by taking inverses if necessary. By Collins' Lemma, we may assume $U \equiv q\ Y_1 U_0$, $V \equiv q\ Y_2 V_0$ where $U_0, V_0 \in J$ and U_0, V_0 do *not* begin with an $s_\alpha^{\pm 1}$. (i.e. they are empty or begin with an $\bar{s}_\alpha^{\pm 1}$). By Collins' Lemma, $U \underset{G}{\sim} V \leftrightarrow \exists W(t_i, \bar{s}_\alpha^{-1} a_\alpha, b_\alpha, c_\alpha$ such that $q\ Y_1 U_0 = W^{-1}\ q\ Y_2 V_0 W$. But by Theorem A and Lemma 18, W cannot contain b_α or c_α. Hence we can write, by Lemmas 16 and 17, $W = \bar{Z}\,\theta$ where θ is a word on t_i, a_α and $\theta^{-1} q\ \theta = \bar{X}_0\ q\ Y_0$ where $X_0 Y_0 \underset{H}{=}$ Thus $W^{-1}\ q\ Y_2 V_0 W \underset{G}{=} \bar{Z}^{-1}\ \bar{X}_0\ q\ Y_0 Y_2 V_0 \bar{Z} \underset{F}{=} q\ Y_1 U_0$. Since this equation holds in F, we conclude $Z = X_0$. By the hypothesis on U_0, V_0 we also have: $Y_0 Y_2 = Y_1$ and $V_0 \bar{Z} = U_0$. So $V_0 U_0^{-1} = \bar{Z}$ must be a word on $\bar{s}_\alpha$ alone and $Y_0 \underset{J}{=} Y_1 Y_2^{-1}$. Since $X_0 = Z$, these equations determine X_0, Y_0 effectively from the given words. Now, in view of Lemma 19, $U \underset{G}{\sim} V \leftrightarrow X_0 Y_0 \underset{H}{=} 1$. Consequently, whether or not $U \underset{G}{\sim} V$ is reducible to the word problem for H.

Case 3 (#): Next assume that U is P-free, but involves an odd number of q-symbols. Taking inverses and permutations as necessary, we may assume the first two q-symbols in U and V are positive (there must be two circularly adjacent q-symbols of the same sign since the number of q-symbols is odd). Using Collins' Lemma we may assume $U \equiv q\ Y_1U_0\overline{X}_1\ q\ M_1$ and $V \equiv q\ Y_2V_0\overline{X}_2\ q\ M_2$ where $U_0, V_0 \in J$ do not begin with $s_i^{\pm 1}$ and do not end with $\overline{s}_i^{\pm 1}$. Again by Collins' Lemma, $U \underset{G}{\sim} V \leftrightarrow \exists W(t_i, \overline{s}_\alpha^{-1} a_\alpha, b_\alpha, c_\alpha)$ such that $U \underset{G}{=} W^{-1}\ V\ W$. As in Case 2, by Theorem A and Lemma 18, W cannot contain a b_α or c_α since the number of q-symbols is odd. Now we write $W = \overline{Z}\theta$ where θ is a word on t_i and a_α and $\theta^{-1}\ q\ \theta = \overline{X}_0\ q\ Y_0$ where $X_0Y_0 \underset{H}{=} 1$. Assuming $U \underset{G}{\sim} V$, we have

$$\begin{aligned} W^{-1}\ V\ W &= \theta^{-1}\overline{Z}^{-1}\ q\ Y_2V_0\overline{X}_2\ q\ M_2\overline{Z}\theta \\ &= \overline{Z}^{-1}\overline{X}_0\ q\ Y_0Y_2V_0\overline{X}_2\overline{X}_0\ q\ Y_0M_2(\overline{X}_0qY_0,\tilde{s})\overline{Z} \\ &\underset{F}{=} q\ Y_1U_0\overline{X}_1\ q\ M_1. \end{aligned}$$

Note that no q-symbols are cancelled because V is q-cyclically reduced and so cannot be conjugate to any word with fewer q-symbols. Since the last equality holds in F, it follows that $Z = X_0$ and $Y_1U_0\overline{X}_1 = Y_0Y_2V_0\overline{X}_2\overline{X}_0$ so by the hypothesis on U_0, V_0 we have $Y_0 = Y_1Y_2^{-1}$, $U_0 = V_0$, and $\overline{X}_0 = \overline{X}_2^{-1}\overline{X}_1$. Now X_0, Y_0 and hence Z are effectively determined from these equations. Moreover, $X_0Y_0 \underset{H}{=} 1$. Now θ may not be uniquely determined, but $\overline{Z}$ is uniquely determined. Let θ' be another such θ and $W' = \overline{Z}\theta'$. Since $\theta'^{-1}\ q\ \theta' = \overline{X}_0\ q\ Y_0$ and $\theta'^{-1}\ \tilde{s}\ \theta' = \tilde{s}$, $W'^{-1}\ V\ W' = U$ if and only if $W^{-1}VW = U$. Hence $U \underset{G}{\sim} V \leftrightarrow X_0Y_0 \underset{H}{=} 1$ where X_0, Y_0 are effectively determined from U and V *and* for $\overline{Z} = \overline{X}_0$ and $W = \overline{Z}\theta$ where $\theta^{-1}\ q\ \theta = \overline{X}_0\ q\ Y_0$ as above we have $U = W^{-1}\ V\ W$. Thus the question as to whether or not $U \underset{G}{\sim} V$ is reducible to the word problem for H.

Case 4: Assume that U is P-free, but involves an even number of q-symbols. By Collins' Lemma, taking inverses and permutations as necessary, we can assume

$$U \equiv q\ U_1\ q^{\varepsilon_2} U_2 \ldots q^{\varepsilon_{2k}} U_{2k} \text{ and } V \equiv q\ V_1\ q^{\varepsilon_2} V_2 \ldots q^{\varepsilon_{2k}} V_{2k},$$

where $\varepsilon_2 = +1$ if there are two circularly adjacent q-symbols and otherwise the q-symbols have alternating signs. Now by Collins Lemma, $U \underset{G}{\sim} V \leftrightarrow \exists\, W(t_i, \overline{s}_\alpha^{-1} a_\alpha, b_\alpha, c_\alpha)$ such that $W^{-1} V W = U$. Writing $W = \overline{Z}\ \theta$ this equation becomes $U(q,\tilde{s}) = \theta^{-1}\overline{Z}^{-1}\ V(q,\tilde{s})\overline{Z}\ \theta = \overline{Z}^{-1}\ V(\overline{X}\ q\ Y,\tilde{s})\overline{Z}$ and in more detail

$$U \underset{F}{=} \overline{Z}^{-1}\ \overline{X}\ q\ YV_1(\overline{X}qY)^{\varepsilon_2} V_2 \ldots (\overline{X}qY)^{\varepsilon_{2k}} V_{2k}\ \overline{Z},$$

where θ is a word of L. By Lemma 21, using the fact that there are an even number of q-symbols, $U \underset{G}{\sim} V \leftrightarrow \exists X, Y$ such that

$$U \underset{F}{=} q\ YV_1(\overline{X}\ q\ Y)^{\varepsilon_2} V_2 \ldots (\overline{X}\ q\ Y)^{\varepsilon_{2k}} V_{2k}\ \overline{X},$$

since if such X, Y exist $W = \overline{Z}\ \theta$ as above can be effectively found so that θ is a word on the a_α, b_α, and c_α alone. In case $\varepsilon_2 = +1$, this last equation determines X, Y uniquely and effectively, since then $U_1 \underset{J}{=} Y\ V_1\ \overline{X}$. Writing $U_1 \equiv Y_1 U_0 \overline{X}_1$ and $V_1 = Y_2 V_0 \overline{X}_2$, where U_0, V_0 do not begin with $s_\alpha^{\pm 1}$ and do not end with $\overline{s}_\alpha^{\pm 1}$ as in Case 3. Now

$U_1 \equiv Y_1 U_0 \overline{X}_1 \underset{J}{=} YV_1\overline{X} = YY_2V_0\overline{X}_2\overline{X}$ implies, by the hypothesis on U_0 and V_0, that $Y = Y_1 Y_2^{-1}$ and $\overline{X} = \overline{X}_2^{-1}\overline{X}_1$ so that X, Y are effectively determined by U and V. Hence the existence of such X and Y is decidable. Finally, assuming $\varepsilon_2 = -1$ and the q-symbols have alternating signs one has the equations:

$$Y\, V_i\, Y^{-1} \underset{J}{=} U_i \qquad \text{for } i \text{ odd}$$

$$\overline{X}^{-1}\, V_i \overline{X} \underset{J}{=} U_i \qquad \text{for } i \text{ even.}$$

Remembering the conventions on X, Y notation, it is an elementary exercise to decide whether or not these equations can be satisfied in the free group J and if so to find such an X and Y.

Hence, whether or not such X, Y exist and therefore whether or not $U \underset{G}{\sim} V$ is decidable. Note that the possibility of conjugating by b_α and c_α has eliminated via Lemma 21 the added requirement that $XY \underset{H}{=} 1$ which occurred in Cases 2 and 3. This is the only reason for introducing the b_α and c_α into the construction. Observe that the $\overline{X}$, Y found in case $\varepsilon_2 = -1$ are not necessarily unique thus causing difficulty had one needed to have $XY \underset{H}{=} 1$.

Case 5: Assume that U is q-free, but involves P-symbols. By Lemma 15 and Collins' Lemma, it may be assumed that $U = U_0\, \theta_1$ and $V = V_0\, \theta_2$ where $U_0, V_0 \in J$ and $\theta_1, \theta_2 \in L$ and are P-parallel. The group K obviously has solvable conjugacy problem. First test whether or not $U \underset{K}{\sim} V$. If so, then $U \underset{G}{\sim} V$ and we are finished. Now suppose $U \underset{K}{\not\sim} V$. In view of Collins' Lemma, we can assume $U \underset{G}{\sim} V \leftrightarrow \exists W \in F$ such that $U = W^{-1}\, V\, W$. By the above reduction, it follows that W involves q since otherwise $U \underset{K}{\sim} V$. Moreover, we assume W involves a minimum number of q-symbols among all such $W \in F$.

From the minimality assumption it follows that W involves exactly one q-symbol. Now by hypothesis W involves at least one q-symbol, so suppose W involves two q-symbols and is q-reduced. Write $W \equiv E_1 q^{\varepsilon} E_2 q^{\delta} W_1$ where $E_1, E_2 \in J$. From the equation $U = W^{-1}\ V\ W$ or $U = W_1^{-1} q^{-\delta} E_2^{-1} q^{-\varepsilon} E_1^{-1}\ V\ E_1 q^{\varepsilon} E_2 q^{\delta} W_1$ since U, V are q-free, it follows that the q-symbols in W must pinch across the intervening word. Assume say that $\varepsilon = +1$, $\delta = -1$. Then

$$E_1^{-1}\ V\ E_1 \underset{G}{=} \theta_2\ (t_i,\ \bar{s}_\alpha^{\ -1} a_\alpha,\ b_\alpha,\ c_\alpha)$$

and

$$q^{-1} E_1^{-1}\ V\ E_1\ q \underset{G}{=} \theta_2\ (t_i R_i^{-1},\ a_\alpha s_\alpha^{-1},\ c_\alpha s_\alpha^{-1},\ b_\alpha s_\alpha^{-1})$$

and

$$E_2^{-1} q^{-1} E_1^{-1}\ V\ E_1 q E_2 \underset{G}{=} \theta_2 (t_i R_i^{-1},\ a_\alpha s_\alpha^{-1},\ c_\alpha s_\alpha^{-1},\ b_\alpha s_\alpha^{-1})$$

and

$$q\ E_2^{-1} q^{-1} E_1^{-1}\ V\ E_1 q E_2 q^{-1} \underset{G}{=} \theta_2 (t_i,\ \bar{s}_\alpha^{-1}\ a_\alpha,\ b_\alpha,\ c_\alpha).$$

Hence $U \underset{G}{=} W_1^{-1} E_1^{-1}\ V\ E_1 W_1$ so that U and V are conjugate by $E_1 W_1$ which involves two fewer q-symbols than W. The dual analysis for the case $\varepsilon = -1$, $\delta = +1$ yields the same result.

Next assume $\varepsilon = +1$, $\delta = +1$. Then

$$E_1^{-1}\ V\ E_1 \underset{G}{=} \theta_2 (t_i,\ \bar{s}_\alpha^{-1}\ a_\alpha,\ b_\alpha,\ c_\alpha)$$

and

$$q^{-1} E_1^{-1}\ V\ E_1 q \underset{G}{=} \theta_2 (t_i R_i^{-1},\ a_\alpha s_\alpha^{-1},\ c_\alpha s_\alpha^{-1},\ b_\alpha s_\alpha^{-1})$$

and

$$E_2^{-1} q^{-1} E_1^{-1}\ V\ E_1 q E_2 \underset{G}{=} \theta_2 (t_i,\ \bar{s}_\alpha^{-1}\ a_\alpha,\ c_\alpha,\ b_\alpha)$$

and

$$q^{-1}E_2^{-1}q^{-1}E_1^{-1}\ VE_1qE_2q \underset{G}{=} \theta_2(t_iR_i^{-1},\ a_\alpha s_\alpha^{-1},\ b_\alpha s_\alpha^{-1},\ c_\alpha s_\alpha^{-1}).$$

But from these equations, it follows that

$$E_2^{-1}\ \theta_2(t_iR_i^{-1},\ a_\alpha s_\alpha^{-1},\ c_\alpha s_\alpha^{-1},\ b_\alpha s_\alpha^{-1})\ E_2 \underset{K}{=} \theta_2(t_i,\ \bar{s}_\alpha^{-1}a_\alpha,\ c_\alpha,\ b_\alpha).$$

Consequently, using the relations of K,

$$E_2E_1^{-1}\ VE_1E_2^{-1} \underset{K}{=} E_2\theta_2(t_i,\ \bar{s}_\alpha^{-1}a_\alpha,\ b_\alpha,\ c_\alpha)E_2^{-1}$$

$$\underset{K}{=} \theta_2(t_iR_i^{-1},\ a_\alpha s_\alpha^{-1},\ b_\alpha s_\alpha^{-1},\ c_\alpha s_\alpha^{-1}).$$

Hence $U \underset{G}{=} W_1^{-1}E_2E_1^{-1}\ V\ E_1E_2^{-1}W_1$, so that U and V are conjugate by $E_1E_2^{-1}W_1$ which involves two fewer q-symbols than W. The dual analysis for the case $\varepsilon = -1$, $\delta = -1$ yields the same result. This justifies the claim that W involves exactly one q-symbol.

In view of the above, $U \underset{G}{\sim} V \leftrightarrow \exists\ E_1,\ E_2 \in J$ such that $U = W^{-1}\ V\ W$ where $W = E_1q^\varepsilon E_2$. From this we get the equations

$$U = E_2^{-1}q^{-\varepsilon}E_1^{-1}\ V\ E_1q^\varepsilon E_2$$

and

$$E_2\ U\ E_2^{-1} \underset{G}{=} q^{-\varepsilon}\ E_1^{-1}\ V\ E_1q^\varepsilon$$

In case $\varepsilon = +1$ this implies

$$E_1^{-1}\ V\ E_1 \underset{K}{=} \theta_2(t_i,\ \bar{s}_\alpha^{-1}a_\alpha,\ b_\alpha,\ c_\alpha)$$

and

$$E_2\ UE_2^{-1} \underset{K}{=} \theta_2(t_iR_i^{-1},\ a_\alpha s_\alpha^{-1},\ b_\alpha s_\alpha^{-1},\ c_\alpha s_\alpha^{-1})$$

and $q^{-1}E_1^{-1}\ V\ E_1q \underset{G}{=} E_2\ U\ E_2^{-1}$. But whether or not such $E_1,\ E_2 \in J$ exist can be effectively determined from the given words using the solution to the conjugacy problem in K. The dual analysis in case $\varepsilon = -1$ yields the same conclusion. Hence whether or not $U \underset{G}{\sim} V$ is decidable.

Case 6: Assume that U involves P-symbols and an even number of q-symbols. In view of the defining relations for G, Lemma 16, and Collins' Lemma, we may suppose that $U \equiv q\ U_0\ \theta_1$, $V \equiv q\ V_0\ \theta_2$ where $U_0, V_0 \in F$ and $\theta_1, \theta_2 \in L$ and are P-parallel. Moreover, we may suppose by Collins' Lemma that $U \underset{G}{\sim} V \leftrightarrow \exists W(t_i, \bar{s}_\alpha^{-1} a_\alpha, b_\alpha, c_\alpha)$ such that $U \underset{G}{=} W^{-1} V W$. Assume temporarily that such a W exists. Write $W = \bar{Z}\theta_0$, $qU_0 = U_1(q,\tilde{s})$, and $qV_0 = V_1(q,\tilde{s})$. For some $\bar{X}, Y$ we have $\theta_0^{-1} V,(q,\tilde{s})\theta_0 = V_1(\bar{X}qY,\tilde{s})$. Hence

$$U = U_1(q,\tilde{s})\theta_1 = W^{-1}\ V_1(q,\tilde{s})\theta_2\ W$$

$$(\ddagger) \qquad = \theta_0^{-1}\ \bar{Z}^{-1}\ V_1(q,\tilde{s})\theta_2\ \bar{Z}\ \theta_0$$

$$= \bar{Z}^{-1}\ V_1(\bar{X}qY,\tilde{s})\bar{Z}\ \theta_0^{-1}\ \theta_2\ \theta_0$$

where in the last step one has $\theta_0^{-1}\ \theta_2\ \theta_0$ since V_1 involves an even number of q-symbols. Hence $\theta_1 \underset{L}{=} \theta_0^{-1}\ \theta_2\ \theta_0$ and $U_1(q,\tilde{s}) \underset{F}{=} \bar{Z}^{-1}V_1(\bar{X}qY$ Writing $\theta_0 \equiv \theta_3\ \theta_4$ for some θ_3 of length less than θ_2 in L, we must have $\theta_3^{-1}\ \theta_2\ \theta_3 = \theta_1$ and $\theta_4^{-1}\ \theta_1\ \theta_4 \underset{L}{=} \theta_1$. Thus replacing V by $W_3^{-1}\ V\ W_3$ where $W_3 = \theta_3(t_i, \bar{s}_\alpha^{-1} a_\alpha, b_\alpha, c_\alpha)$ and moving $\tilde{s}$-symbols to the left by conjugation, we may assume $\theta_1 = \theta_2$ in our original equation. Note that there is an effective method for obtaining such a new V from the original (assuming this is possible which can also be tested). Changing notation, we may suppose

$$U = qU_0\theta_1 = U_1(q,\tilde{s})\theta_1,\ V = qV_0\theta_1 = V_1(q,\tilde{s})\theta_1$$

and that $U \underset{G}{\sim} V \leftrightarrow \exists\, W(t_i, \bar{s}_\alpha^{-1} a_\alpha, b_\alpha, c_\alpha)$ such that $U \underset{G}{=} W^{-1} V W$

where $W = \bar{Z}\theta_0$ and $\theta_0^{-1}\theta_1\theta_0 \underset{L}{=} \theta_1$. Write $\theta_1 = \Gamma^m$ for m maximal.

Then Γ is not a proper power and can be effectively determined from θ_1.

The equation $\theta_0^{-1}\theta_1\theta_0 \underset{L}{=} \theta_1$ implies (since L is free) that $\theta_0 = \Gamma^k$ for

some k. Now $W = \bar{Z}\theta_0$ implies $W = [\Gamma(t_i, \bar{s}_\alpha^{-1} a_\alpha, b_\alpha, c_\alpha)]^k$. So

$U \underset{G}{\sim} V \leftrightarrow \exists\, k$ such that $W^{-1} V W \underset{G}{=} U$ where $W = [\Gamma(t_i, \bar{s}_\alpha^{-1} a_\alpha, b_\alpha, c_\alpha)]^k$

and Γ is effectively determined by θ_1 and hence by U and V. Writing

$\Gamma(t_i, \bar{s}_\alpha^{-1} a_\alpha, b_\alpha, c_\alpha) = \bar{Z}_0\, \Gamma$, it follows that

$$\begin{aligned} W^{-1} V W &= \Gamma^{-k}\, \bar{Z}_0^{-k}\, V_1(q, \tilde{s})\theta_1\, \bar{Z}_0^k\, \Gamma^k \\ &= \bar{Z}_0^{-k}\, V_1(\bar{X}^k\, q\, Y^k, \tilde{s})\bar{Z}_0^k\, \Gamma^{-k}\, \theta_1\, \Gamma^k \\ &= \bar{Z}_0^{-k}\, V_1(\bar{X}^k\, q\, Y^k, \tilde{s})\bar{Z}_0^k\, \theta_1 \end{aligned}$$

where $\bar{Z}_0$, $\bar{X}$, Y can be effectively determined from Γ (just conjugate

V_1 by $\Gamma^{\pm 1}$ to find $\bar{X}$, Y). Thus $U \underset{G}{\sim} V \leftrightarrow \exists\, k$ such that

$$U_1(q, \tilde{s}) \underset{F}{=} \bar{Z}_0^{-k} V_1(\bar{X}^k q Y^k, \tilde{s})\, \bar{Z}_0^k$$

where Z_0, X, Y can be effectively found from the given words U and V. But whether or not such a k exists is a solvable problem. This is easily established by a direct argument since the equation holds in the free group F. Alternatively, the solvability follows from the main result of Lyndon [28] since one can effectively decide whether a polynomial in a single variable has an integer solution.

Case 7: Finally assume that U involves P-symbols and an odd number of q-symbols. As in Case 6, we may suppose $U \equiv qU_0\theta_1$, $V \equiv qV_0\theta_2$ where U_0, $V_0 \in F$ and $\theta_1, \theta_2 \in L$ and are P-parallel. If θ_1 (and hence θ_2) involve *neither* b_α *nor* c_α, then the argument given in Case 6 also applies to Case 7. This will become clear in what follows.

Again by Collins' Lemma, we can suppose that

$$U \underset{G}{\sim} V \leftrightarrow \exists W(t_i, \bar{s}_\alpha^{-1} a_\alpha, b_\alpha, c_\alpha)$$

such that $U \underset{G}{=} W^{-1} V W$. Assume temporarily that such a W exists. While $W = \bar{Z}\theta_0$, $qU_0 = U_1(q,\tilde{s})$, and $q\ V_0 = V_1(q,\tilde{s})$. For some $\bar{X}$, Y we have $\theta_0^{-1} V_1(q,\tilde{s})\ \Gamma_0 = V_1(\bar{X}qY,\tilde{s})$, where for $\theta_0 = \theta_0(t_i, a_\alpha, b_\alpha, c_\alpha)$, Γ_0 is $\theta_0(t_i, a_\alpha, c_\alpha, b_\alpha)$. Hence, by analogy to ($\ddagger$),

$$\begin{aligned} U = U_1(q,\tilde{s})\theta_1 &= W^{-1} V_1(q,\tilde{s})\theta_2 W \\ (*) \qquad &= \theta_0^{-1}\bar{Z}^{-1} V_1(q,\tilde{s})\theta_2 \bar{Z}\theta_0 \\ &= \bar{Z}^{-1} V_1(\bar{X}qY,\tilde{s})\bar{Z}\ \Gamma_0^{-1}\ \theta_2\,\theta_0. \end{aligned}$$

The equation $\theta_0^{-1} V_1(q,\tilde{s})\ \Gamma_0 = V_1(\bar{X}qY,\tilde{s})$ follows from the fact that V_1 involves an odd number of q-symbols. Since $U \equiv U_1(q,\tilde{s})\theta_1$, from (*) it follows that $\theta_1 = \Gamma_0^{-1}\ \theta_2\,\theta_0$.

In case θ_1 (and hence θ_2) involve *neither* b_α *nor* c_α, because of the relationship between θ_0 and Γ_0 it follows that θ_0 (and hence Γ_0) involve neither b_α nor c_α. Hence, in this case, Γ_0 is θ_0 and the equation ($\ddagger$) of Case 6 holds. Consequently, the algorithm given in Case 6 also works for Case 7 when θ_1 and θ_2 involve neither b_α nor c_α. Thus, for the remainder of Case 7, we may assume that θ_1 (and hence θ_2) involve a b_α or a c_α

From the equation $\theta_1 = \Gamma_0^{-1}\ \theta_2\theta_0$ and the fact that V is P-cyclically reduced, it follows that either $\Gamma_0^{-1}\ \theta_2$ or $\theta_2\theta_0$ is reduced without cancellation in L. Since U, V are P-cyclically reduced and length θ_1 = length θ_2 as elements of the free group L, conjugation of V by θ_0 is equivalent to a sequence of moves of exactly one of the following types: (1) Move the final P-symbol in θ_2 to the front of V, then pass this symbol through $V_1(q,\tilde{s})$ to get $V_1'(q,\tilde{s})\theta_2'$; (2) Move the initial P-symbol in θ_2 left through $V_1(q,\tilde{s})$ then from the front to the end of the resulting word thus obtaining $V_1'(q,\tilde{s})\theta_2'$. Assuming θ_0 is freely reduced, then $\theta_0^{-1}\ V_1\theta_2\theta_0$ can be obtained from $V_1\theta_2 = V$ by a sequence of all type (1) or all type (2) moves. Let $\theta_2, \theta_2', \ldots, \theta_2^{(k)}$ be the resulting sequence of P-parts at any stage. Then it is clear that this sequence has period at most 2 times length θ_2. Consequently, after at most 2 times length θ_2 conjugations of V by P-symbols, the resulting P-part must be θ_1. Thus for suitable θ_3 in L of length at most 2 times length θ_2, $\theta_3^{-1}\ V\ \theta_3 = \theta_3^{-1}\ V_1\theta_2\theta_3$ has P-part θ_1 after P-reduction. Consequently, replacing V by $W_3^{-1}\ V\ W_3$ where

$$W_3 = \theta_3(t_i,\ \bar{s}_\alpha^{-1}\ a_\alpha,\ b_\alpha,\ c_\alpha)$$

we may assume that θ_2 is θ_1 in the equations (*). (Note that one can effectively tell whether such θ_3 exists and if so find it. If no such θ_3 exists then $U \underset{G}{\not\sim} V$ by the above argument).

If $\theta_j = \theta_j(t_i,\ a_\alpha,\ b_\alpha,\ c_\alpha)$ we write Γ_j for $\theta_j(t_i,\ a_\alpha,\ c_\alpha,\ b_\alpha)$, with b_α and c_α interchanged. By the above reductions, it follows that $\theta_1 \underset{L}{=} \Gamma_0^{-1}\ \theta_1\theta_0$ where we are assuming θ_1 involves either b_α or c_α. By the relationship between Γ_j and θ_j, and the fact that L is a free group and either $\Gamma_0^{-1}\ \theta_1$ or $\theta_1\theta_0$ must be reduced without cancellation, it follows that either (1) θ_0 is $[\theta_1^{-1}\ \Gamma_1^{-1}]^k$ and $\Gamma_0^{-1}$is $[\theta_1\ \Gamma_1]^k$ for some $k \geq 0$; or (2) θ_0 is $[\Gamma_1\ \theta_1]^k$ and Γ_0^{-1} is $[\Gamma_1^{-1}\theta_1^{-1}]^k$ for some $k \geq 0$. (Note that Γ_1 and θ_1 are *not* the same by hypothesis.) In any event θ_0 is $[\Gamma_1\theta_1]^k$ for some k (positive or negative).

Now for $\overline{X}_0$, Y_0 which can be effectively found from θ_1, we must have

$$\theta_0^{-1}\ V_1(q,\tilde{s})\ \Gamma_0 = V_1(\overline{X}_0^{\,k}\ q\ Y_0^{\,k},\ \tilde{s});\ \text{here}$$

$$[\Gamma_1\theta_1]^{-1}\ V_1(q,\tilde{s})[\theta_1\ \Gamma_1] = V_1(\overline{X}_0\ q\ Y_0,\tilde{s}).$$

Thus, after completing effectively all of the above reductions, it follows that $U \underset{G}{\sim} V \leftrightarrow \exists\, k$ such that $U_1(q,\tilde{s}) \underset{F}{=} \overline{Z}_0^{\,-k}\ V_1(\overline{X}_0^{\,k} q Y_0^{\,k},\tilde{s})\,\overline{Z}_0^{\,k}$ where X_0, Y_0 are as above and $\overline{Z}_0$ is the $\overline{s}$-part of

$$\Gamma_1(t_i,\ \overline{s}_\alpha^{-1} a_\alpha,\ b_\alpha,\ c_\alpha)\theta_1(t_i,\ \overline{s}_\alpha^{-1} a_\alpha,\ b_\alpha,\ c_\alpha).$$

(See Case 6 for more details on this: recall that in $W = \overline{Z}\,\theta_0$, $\overline{Z}$ is determined as above from θ_0). Now by the argument given in Case 6, this is a solvable problem. Hence, in Case 7, whether or not $U \underset{G}{\sim} V$ is decidable.

Since every (P,q)-cyclically reduced word of G belongs to one of the classes considered in Cases 1-7 and one can effectively tell which class (by inspection), this completes the proof of Theorem 22.||

THEOREM 24: G *is residually finite.*

Proof: G is a Britton extension of F by the stable letters of P. Let N be the fully invariant subgroup of F generated by all squares (i.e. F/N satisfies the relation $W^2 = 1$ identically). Now $N \lhd G$ since each word of N contains an even number of q-symbols so that $b_i^{-1} N b_i = N$ and $c_i^{-1} N c_i = N$ (the same for t_i, a_α is obvious). But N is free and f.g. since F/N is finite (even an abelian 2-group). By Theorem III-7, to show G residually finite it is enough to show G/N is residually finite.

Now $L \cap N = 1$ so L is naturally embedded in G/N. Let A be the subgroup of G/N generated by the (images of the) $\tilde{s}$. Then A is a finite abelian 2-group and, in view of the relations for G itself, $L \times A \subset G/N$. Moreover q commutes with $\tilde{s}$ in G/N and conjugation by q sends elements of L to elements of $L \times A$. Since $L \times A$ and q generate G/N, it follows that $L \times A \lhd G/N$. Now, $G/N/L \times A = \langle q; q^2 = 1 \rangle$, a finite group. $L \times A$ is clearly f.g. residually finite. Hence by [38] Theorem 26.12 (due to Gruenberg), G/N is residually finite as was to be proved. (Actually G/N is easily seen to be a split extension of $L \times A$ by $\langle q; q^2 = 1 \rangle$ so Theorem III-7 applies). ||

This also completes the proof of Theorem 12.

C. SOME APPLICATIONS

The following result strengthens a theorem of Collins [19] for recursively presented groups.

THEOREM 25: *The property of having a solvable conjugacy problem is not hereditary in the class of f.p. residually finite groups. In particular, there is a f.p. residually finite group* B *with solvable conjugacy problem which has a f.p. subgroup* G *with conjugacy problem of arbitrary r.e. degree* D.

Proof: Take G to be the group constructed for Theorem 12 above. Let B be the f.p. group defined as follows:

Generators: Generators of G, $t_{m+1}, \ldots, t_{m+n}$

Relations: Relations of G and the relations

$$t_j^{-1}\ q\ t_j = qs_{j-m} \qquad m+1 \leq j \leq m+n$$

$$t_j^{-1}\ \tilde{s}\ t_j = \tilde{s} \qquad \text{each } \tilde{s}.$$

Clearly B is the group obtained by applying our construction of G from H to the group $H_0 = < s_1,\dots,s_n;\ R_1,\dots,R_m,\ s_1,\dots,s_n >$. But H_0 is trivial and hence has solvable word problem. By Theorem 12, B is residually finite and has solvable conjugacy problem. But obviously G is embedded in B by the natural map. Thus Theorem 12 gives the desired result.||

The following application of Theorem 12 answers a question of Boone:

THEOREM 26: *Let* E *be a group with unsolvable word problem. Then there is a recursive class* $\theta = \{G_W,\ W \epsilon E\}$ *of finite presentations of groups indexed by words of* E *such that: (1) each* G_W *is residually finite; and (2)* $G_W \cong G_1 \leftrightarrow W = 1$ *in* E. *From (2), it follows that the isomorphism problem for* θ *is unsolvable. In view of (1), each* $G_W \epsilon \theta$ *has a solvable word problem and the word problems for various* $G_W \epsilon \theta$ *are solvable by a uniform algorithm.*

REMARK: In the usual proofs that the isomorphism problem is unsolvable, the indexing group E is usually embedded in members of the class corresponding to $W \neq 1$ in E. Consequently, such groups have unsolvable word problem.

Proof: The reader is assumed to be familiar with the construction of Rabin [41] or with Section V-C of this work. By Rabin's Theorem, there is a recursive class $\Gamma = \{H_W,\ W \epsilon E\}$ such that H_W has solvable word problem $\leftrightarrow W = 1$, in E, since "has solvable word problem" is a Markov property. From the proof of Rabin's Theorem, it may be assumed that each H_W has presentation of the form $< s_1,\dots,s_n;\ Ws_1^{-1},\ R_2,\dots,R_m >$ where W is a word on the s_α and the presentations for varying H_W vary only in that the relation $Ws_1^{-1} \equiv R_1$ changes; i.e. n, m and $R_2,\ldots,R_m$ are the same for all H_W. (In Section V-C, take $r = s_1$ and replace $[W,x]$ by W since $W = 1 \leftrightarrow [W,x] = 1$ in the construction). Moreover, from the proof of Rabin's Theorem it follows that $W = 1$ in $H_W \leftrightarrow W = 1$ is a consequence of the relations $R_2,\ldots,R_m$ (note *not* $R_1 \equiv Ws_1^{-1}$).

Let G_W be the result of applying the construction for Theorem 12 to H_W. Then varying G_W differ only in one relation, namely

$$t_1^{-1}\, q\, t_1 = q\, Ws_1^{-1}.$$

Each G_W has the same generators, and all other defining relations the same. By Theorem 12, each G_W is residually finite. Clearly $\theta = \{G_W,\ W \in E\}$ is a recursive class of finite presentations of groups. By Theorem 12, G_1 has solvable conjugacy problem since H_1 has solvable word problem. If $W \neq 1$ in E, then H_W has unsolvable word problem and by Theorem 12 G_W has unsolvable conjugacy problem. Thus $W \neq 1$ in $E \rightarrow G_W \ncong G_1$.

Assume $W = 1$ in E. It remains to show that $G_W \cong G_1$. By Lemma 19 and its proof and the fact that $W = 1$ in H_W is a consequence of $R_2,\ldots,R_m$, it follows that there is a word $U = U(t_i, a_\alpha)$ where only t_i $(i \geq 2)$ appear in U such that $U^{-1}\, q\, U = qW$ (and of course $U^{-1}\tilde{s}U = \tilde{s}$). Moreover, this equation holds in every G_W since only relations not involving t_1 are necessary in the proof of the equation.

For simplicity, let z be a variable ranging over the generators of G_W

except t_1. Define a map $\phi: G_W \to G_1$ by the equations $\phi(z) = z$ and $\phi(t_1) = t_1U$. We must show ϕ defines a homomorphism. It is enough to show $\phi(t_1^{-1}\tilde{s}\,t_1\,\tilde{s}^{-1}) = 1$ in G_1 and $\phi(t_1^{-1}\,q\,t_1\,s_1\,W^{-1}q^{-1}) = 1$ in G_1 since all other relations involve only z-generators and $\phi(z) = z$. Now in G_1: $\phi(t_1^{-1}\tilde{s}t_1\tilde{s}^{-1}) = U^{-1}t_1^{-1}\tilde{s}t_1U\tilde{s}^{-1} = 1$ since $t_1^{-1}\tilde{s}t_1 = \tilde{s}$ and $U^{-1}\tilde{s}U = \tilde{s}$ Also

$$\phi(t_1^{-1}qt_1s_1W^{-1}q^{-1}) = U^{-1}t_1^{-1}qt_1Us_1W^{-1}q^{-1}$$

$$\underset{G_1}{=} U^{-1}qs_1^{-1}Us_1W^{-1}q^{-1}$$

$$= U^{-1}qUW^{-1}q^{-1} = qW\,W^{-1}q^{-1} = 1.$$

Thus ϕ defines a homomorphism. Since U does not involve t_1, it is clear that ϕ is onto.

Next define ϕ^{-1}: $G_1 \to G_W$ (which we will verify is in fact inverse to ϕ) by the equations $\phi^{-1}(z) = z$ and $\phi^{-1}(t_1) = t_1U^{-1}$. Then $\phi^{-1}(t_1^{-1}\,\tilde{s}\,t_1\,\tilde{s}^{-1}) = 1$ in G_W as before and in G_W one has

$$\phi^{-1}(t_1^{-1}\,q\,t_1\,s_1\,q^{-1}) = Ut_1^{-1}\,qt_1\,U^{-1}\,s_1\,q^{-1}$$

$$\underset{G_W}{=} U\,q\,W\,s_1^{-1}\,U^{-1}\,s_1\,q^{-1}$$

$$= U\,q\,W\,U^{-1}\,q^{-1} = q\,q^{-1} = 1.$$

Hence ϕ^{-1} is a homomorphism and, as before, ϕ^{-1} is onto. Now $\phi \circ \phi^{-1}(z) = \phi(z) = z$, $\phi^{-1} \circ \phi(z) = \phi^{-1}(z) = z$ and $\phi \circ \phi^{-1}(t_1) = \phi(t_1U^{-1}) = \phi(t_1)\phi(U^{-1}) = t_1\,U\,U^{-1} = t_1$ and $\phi^{-1} \circ \phi(t_1) = \phi^{-1}(t_1U)$ $\phi^{-1}(t_1)\phi^{-1}(U) = t_1U^{-1}U = t_1$.

Thus ϕ^{-1} is inverse to ϕ and so ϕ is an isomorphism. Hence $W = 1$ in $E \to G_W \cong G_1$ as claimed.||

Let G_W be as in the proof of Theorem 26. By Lemma 15 and Lemma II-8, there is a f.g. free group F such that $G_W * F$ is the free product with f.g. amalgamation of two free groups. Moreover, the rank of F does not depend on the particular W since each G_W has the same number of stable letters. Define $A_W = G_W * F$. Then each A_W is again residually finite (see [30], p. 417), and we have the following result:

COROLLARY 27: *Let* E *be a group with unsolvable word problem. Then there is a recursive class* $\Omega = \{A_W, W \in E\}$ *of finite presentations of groups indexed by words of* E *such that: (1) each* A_W *is residually finite; (2) each* A_W *is a free product with f.g. amalgamation of two free groups; and (3)* $A_W \cong A_1 \leftrightarrow W = 1$ *in* E. *In particular the isomorphism problem for* Ω *is unsolvable.*||

REMARK: We state without proof that a similar theorem holds if condition (2) is replaced by "A_W is a split extension of one f.g. free group by another". The proof uses methods of section IV-B applied to a particular case of the construction in section III-A.

Finally, we combine Theorem 12 with a result of Boone and Rogers [15] to obtain:

THEOREM 28: *There is a recursive class* θ *of finite presentations of residually finite groups such that the problem of determining for arbitrary* $G \in \theta$ *whether or not* G *has solvable conjugacy problem is exactly maximal in* Σ_3 *in the Kleene-Mostowski heirarchy.*

Proof: In view of Boone and Rogers [15] Theorem 1 and our Theorem 12 and Boone [11], we need only show how to view a finite presentation of a group G as a logical system L so that the theorems of L are just the conjugacy assertions of G.

Let $G = \langle x_1,\ldots,x_n;\ Y_1,\ldots,Y_m \rangle$. Define a well-formed formula of L to be an expression (U,V) where U and V are words on the x_i. Take L to have the single axiom: (1,1). As rules of inference for L take:

$$(W,W) \rightarrow (WU,WU)$$
$$(U,V) \rightarrow (V,U)$$
$$(UW^{-1}W,V) \rightarrow (U,V)$$
$$(U,V) \rightarrow (UW^{-1}W,V)$$
$$(U,V) \rightarrow (UY_i,V) \qquad (1 \leq i \leq m)$$
$$(UY_i,V) \rightarrow (U,V) \qquad (1 \leq i \leq m)$$
$$(UW,V) \rightarrow (WU,V)$$

for any words U, V, W on the x_i. Here $A \rightarrow B$ means from A deduce B. Then (U,V) is a theorem of L if and only if U and V are conjugate in G.||

CHAPTER V

ON THE ISOMORPHISM PROBLEM FOR GROUPS

A. BACKGROUND AND RESULTS

Rabin [41] has shown that the isomorphism problem for finitely presented groups is unsolvable. Given f.p. groups G and H, Rabin gives a construction yielding a recursive class $\Omega = \{\Pi_W, W \in G\}$ of finite presentations of groups indexed by words of G such that $\Pi_W \cong H \leftrightarrow W \underset{G}{=} 1$ (the empty word). If G is chosen as a group with unsolvable word problem, then the isomorphism problem for Ω (and hence for all f.p. groups) is unsolvable, since one cannot decide whether or not $\Pi_W \cong \Pi_1$.

In [12] Boone modified Rabin's construction to give arbitrary recursively enumerable degree results for the isomorphism problem. Given a f.p. torsion-free group G, Boone constructs a recursive class $\Omega = \{\Pi_W, W \in G\}$ of finite presentations of groups indexed by words of G such that $\Pi_U \cong \Pi_V \leftrightarrow$ either (U is V) or ($U \underset{G}{=} 1$ and $V \underset{G}{=} 1$). Boone has asked whether the above constructions can be strengthened to yield, for any G, a class Ω such that $\Pi_U \cong \Pi_V \leftrightarrow U \underset{G}{=} V$. We answer this natural question affirmatively by proving the following:

THEOREM 1: *Let* G *be a given f.p. group. There is an explicit, effective construction (uniform in* G*) which yields a recursive class* $\Omega = \{\Pi_W, W \in G\}$ *of finite presentations of groups indexed by words of* G *such that* $\Pi_U \cong \Pi_V \leftrightarrow U \underset{G}{=} V$. *Moreover, a version of the construction assures that* Π_1 *is a presentation of the trivial group.*

REMARK: This theorem has the following curious consequence: Ω admits a group structure defined by $(\Pi_U) \circ (\Pi_V) = \Pi_{UV}$.

We can generalize Theorem 1 to the following:

THEOREM 2: *Let* E(i,j) *be an equivalence relation on the natural numbers* N. *Then* E *is r.e.* $\leftrightarrow$ *there is a recursive class* $\Omega = \{\Pi_i, i \in N\}$ *of finite presentations of groups such that* $\Pi_i \cong \Pi_j$ *if and only if* E(i,j).

Using this result one can show the following:

COROLLARY 3: *Every r.e. many-one degree of unsolvability contains an isomorphism problem, but the one-one degree of a simple set does not.*

Notice that Corollary 3 would follow from Theorem 1 alone if there were a word problem for a f.p. group in every many-one degree. However, the latter is an open question.

B. THE BASIC CONSTRUCTION

Our plan is to prove Theorem 1 by a direct construction, and then reduce Theorem 2 to this result. Assuring that Π_1 is trivial is postponed until Section C. We begin with an intuitive discussion of the construction.

We are given a f.p. group G and wish to construct a recursive class $\{\Pi_W, W \in G\}$ so that $\Pi_U \cong \Pi_V \leftrightarrow U \underset{G}{=} V$. Rabin's construction has the property (1) $U \underset{G}{=} V \rightarrow \Pi_U \cong \Pi_V$. Somehow we must insure (2) $U \underset{G}{\neq} V \rightarrow \Pi_U \ncong \Pi_V$. Boone's construction does this for torsion-free G by building in finite cycles. However, Boone's construction does not have the property (1). The construction is to be uniform in G so no special properties of G can be used in the proof.

The first step is to replace G by a torsion-free group H which reflects the relationships between words in G. Let $1 = W_0, W_1, W_2, \ldots$ be a list of words of G. For each i,j a word $\phi(i,j)$ of H is defined so that $\phi(i,j)$ has certain properties of the product $W_i W_j^{-1}$ in G. Let f(j) be the least $i \leq j$ such that $W_i \underset{G}{=} W_j$. Then $f(j)$ will be the least i such that $\phi(i,j) \underset{H}{=} 1$. Moreover, $f(\ell) = f(j) \leftrightarrow W_\ell \underset{G}{=} W_j$. (In general, $f(j)$ is not a recursive function).

Next we enlarge H to another torsion free group L. Then Π_{W_j} is formed by adding to L cyclic groups of various prime orders. However, this is done so that all but the first $f(j) - 1$ cycles collapse. One then shows $\Pi_{W_j} \cong \Pi_{W_{f(j)}}$. By properties of f this shows

$$W_\ell \underset{G}{=} W_j \rightarrow \Pi_{W_\ell} \cong \Pi_{W_j},$$

so that property (1) holds. On the other hand, if $W_\ell \underset{G}{\neq} W_j$ then $f(\ell) \neq f(j)$. But then Π_{W_ℓ} and Π_{W_j} have different cyclic prime order subgroups so that $\Pi_{W_\ell} \ncong \Pi_{W_j}$. Hence (2) holds.

This completes the intuitive description of the construction. We now proceed to the actual proof.

Let G be a given f.p. group, and let $1 = W_0, W_1, W_2, \ldots$ be a recursive (possibly proper) subset of the words of G (we insist $W_0 = 1$ the empty word). Call (H, ϕ) a *covering pair* for G (with respect to the list of words $W_0, W_1, W_2, \ldots$) if the following conditions are satisfied:

(i) H is a f.p. torsion free group.

(ii) $\phi(i,j)$ is a recursive function from pairs of words (W_i,W_j) of G to words of H, and the range of ϕ is a recursive subset of the words of H.

(iii) $\phi(i,j) \underset{H}{=} 1 \leftrightarrow W_i \underset{G}{=} W_j$

(iv) $\phi(i,j) \underset{H}{=} \phi(i,\ell) \leftrightarrow W_j \underset{G}{=} W_\ell$.

A key point in the definition of covering pair is that H must be torsion free. If G itself is torsion free, the pair (G,ϕ) where $\phi(i,j) = W_i W_j^{-1}$

is easily seen to be a covering pair for G. Note that G is *not* embedded in H in general.

THEOREM 4: *Let* $G = < s_1,\dots,s_n; R_1,\dots,R_m >$ *be a given f.p. group, with* $1 = W_0, W_1, W_2, \dots$ *a recursive subset of the words of* G. *Then there is an explicit, effective construction (uniform in* G*) which yields a covering pair* (H,ϕ) *for* G.

Proof: Put $F = < s_1,\dots,s_n >$, a free group. Now form

$$J = < s_1,\dots,s_n, a_1,\dots,a_n; a_i^{-1}s_\alpha a_i = s_i^{-1}s_\alpha s_i \ 1 \leq i, \alpha \leq n >$$

By Britton's Lemma, the map $s_\alpha \to s_\alpha$ naturally embeds F in J since the a_i simply induce inner automorphisms of F. Now define the f.p. group H by:

Generators: $s_1,\dots,s_n,a_1,\dots,a_n, k$

Relations: $a_i^{-1}\, s_\alpha\, a_i = s_i^{-1}\, s_\alpha\, s_i \ (1 \leq i, \alpha \leq n)$

$$k^{-1}\, R_j\, k = R_j \qquad 1 \leq j \leq m$$

$$k^{-1}\, a_i\, k = a_i \qquad 1 \leq i \leq n.$$

Again by Britton's Lemma, the map $s_\alpha \to s_\alpha$, $a_i \to a_i$ naturally embeds J in H. Moreover, $F \subset J \subset H$ is a two-stage Britton tower over F and hence H is torsion-free by Corollary II-3. Let A be the subgroup of J generated by the a_i and the R_j. Define $\phi(i,j)$ to be the word

$$k^{-1}\, W_i\, W_j^{-1}\, k\, W_j\, W_i^{-1}$$

of H. It is clear that the pair (H,ϕ) has properties (i) and (ii) of the definition of covering pair. The proof of Theorem 4 will be completed by the following sequence of Lemmas:

LEMMA 5: *Let* U *be a word of* J. *Then* $U \in A \leftrightarrow U \underset{J}{=} ZT$ *where* T *is a word on* a_i *and* Z *is a word on the* s_α *such that* $Z \underset{G}{=} 1$. *Moreover, if* Z *is a word on the* s_α *alone, then* $Z \in A \leftrightarrow Z \underset{G}{=} 1$.

Proof: The second claim follows immediately from the first. For suppose Z_1, Z_2 are words on s_α, T a word on the a_i. Then $Z_1 \underset{J}{=} Z_2 T \leftrightarrow T$ is freely equal to 1, and $Z_1 \underset{J}{=} Z_2$, by Britton's Lemma for J over F. It remains to prove the first claim. $(\leftarrow)$ Since $Z \underset{G}{=} 1$ we know Z is freely equal to a product $\prod_{i=1}^{d} X_i(s_\alpha)^{-1} R_{\ell_i}^{\varepsilon_i} X_i(s_\alpha)$ where $X_i(s_\alpha)$ are words on the s_α. Let $X_i(a_\alpha)$ be the corresponding word on the a_α. Then

$$Z \underset{J}{=} \prod_{i=1}^{d} X_i(a_\alpha)^{-1} R_{\ell_i}^{\varepsilon_i} X_i(a_\alpha)$$

so ZT is equal in J to a word on the R_j and a_α. That is, $ZT \in A$.

$(\rightarrow)$ Assume $U \in A$. Then

$$U \underset{J}{=} T_1 R_{\ell_1}^{\varepsilon_1} T_2 \ldots T_d R_{\ell_d}^{\varepsilon_d} T_{d+1}.$$

By inserting inverse pairs of a_α we can rewrite this as

$$U \underset{J}{=} T_1^{*-1} R_{\ell_1}^{\varepsilon_1} T_1^{*} T_2^{*-1} R_{\ell_2}^{\varepsilon_2} T_2^{*} \ldots T_d^{*-1} R_{\ell_d}^{\varepsilon_d} T_d^{*} T_{d+1}^{*}.$$

Now $T_i^{*-1} R_{\ell_i}^{\varepsilon_i} T_i^{*} \underset{J}{=} X_i^{-1} R_{\ell_i}^{\varepsilon_i} X_i$ where X_i is the word on s_α corresponding to T_i^{*}. Putting $Z = \prod_{i=1}^{d} X_i^{-1} R_{\ell_i}^{\varepsilon_i} X_i$ and $T = T_{d+1}^{*}$ the result follows.||

LEMMA 6: $\phi(i,j) \underset{H}{=} 1 \leftrightarrow W_i \underset{G}{=} W_j$

Proof: By Britton's Lemma and the relations for H, $\phi(i,j) =$ $k^{-1} W_i W_j^{-1} k W_j W_i^{-1} \underset{H}{=} 1 \leftrightarrow W_i W_j^{-1} \in A \leftrightarrow W_i W_j^{-1} \underset{G}{=} 1$ (by Lemma 5) $\leftrightarrow W_i \underset{G}{=} W_j$. ||

LEMMA 7: $\phi(i,j) \underset{H}{=} \phi(i,\ell) \leftrightarrow W_j \underset{G}{=} W_\ell$.

Proof: Now $\phi(i,j) \underset{H}{=} \phi(i,\ell) \leftrightarrow \phi(i,j)\phi(i,\ell)^{-1} \underset{H}{=} 1$.

But $\phi(i,j)\phi(i,\ell)^{-1} = k^{-1} W_i W_j^{-1} k W_j W_i^{-1} W_i W_\ell^{-1} k^{-1} W_\ell W_i^{-1} k$.

$$= k^{-1} W_i W_j^{-1} k W_j W_\ell^{-1} k^{-1} W_\ell W_i^{-1} k.$$

($\leftarrow$) Assume $W_j \underset{G}{=} W_\ell$. Then applying Lemma 5 and the relations for H we have

$$\phi(i,j)\phi(i,\ell)^{-1} = k^{-1} W_i W_j^{-1} k W_j W_\ell^{-1} k^{-1} W_\ell W_i^{-1} k$$

$$= k^{-1} W_i W_j^{-1} W_j W_\ell^{-1} W_\ell W_i^{-1} k$$

$$= k^{-1} k = 1.$$

($\rightarrow$) Assume $\phi(i,j)\phi(i,\ell)^{-1} \underset{H}{=} 1$. Then by Britton's Lemma one of the following must hold:

(1) $W_i W_j^{-1} \in A$, (2) $W_j W_\ell^{-1} \in A$, or (3) $W_\ell W_i^{-1} \in A$.

Now by Lemma 5, (1) occurs $\leftrightarrow W_i W_j^{-1} \underset{G}{=} 1 \leftrightarrow W_i \underset{G}{=} W_j \leftrightarrow \phi(i,j) = 1$ (by Lemma 6) $\leftrightarrow \phi(i,\ell) = 1$ (by hypothesis) $\leftrightarrow W_i \underset{G}{=} W_\ell$ (by Lemma 6). Hence, if (1) occurs, then $W_j \underset{G}{=} W_\ell$. Case (3) is similar to (1). Assume (2) occurs. Then by Lemma 5, $W_j W_\ell^{-1} \underset{G}{=} 1$ or $W_j \underset{G}{=} W_\ell$. ||

Lemmas 6 and 7 complete the proof of Theorem 4. The construction proving Theorem 4 is no longer needed since the definition of covering pair gives an axiomatic characterization of the necessary properties.

Assume that (H, ϕ) is a covering pair for G (with respect to $1 = W_0, W_1, W_2, \ldots$). Put $L = H * \langle x,y \rangle$, the free product of H and the free group on x and y. The *length* of a word $W \in L$ means the normal form length of W in the free product L. Note that L is torsion-free.

NOTATION: Let $f(\ell)$ = the least $j \leq \ell$ such that $W_j \underset{G}{=} W_\ell$. (Recall that f is not in general a recursive function.) By property (iii) of a covering pair, $f(\ell)$ = the least $j \leq \ell$ such that $\phi(j,\ell) \underset{L}{=} 1$. Consequently, none of $\phi(0,\ell), \phi(1,\ell), \ldots, \phi(f(\ell) - 1, \ell)$ is equal to 1 in L.

The notation $[a,b]$ means the commutator $a^{-1}b^{-1}ab$. Now define

$$\gamma(0,\ell) = [x^{-1}\, y\, x, \phi(0,\ell)]$$

and

$$\gamma(i,\ell) = [[\gamma(i-1),\ell), x^{-(i+1)}y^{i+1}x^{i+1}], \phi(i,\ell)]$$

for $i > 0$.

Observe that $\gamma(i-1,\ell) \underset{L}{=} 1$ implies that $\gamma(i,\ell) \underset{L}{=} 1$. Since $\phi(f(\ell),\ell) \underset{L}{=} 1$, it follows that $\gamma(i,\ell) \underset{L}{=} 1$ for all $i \geq f(\ell)$.

THEOREM 8: $\gamma(0,\ell), \ldots, \gamma(f(\ell) - 1, \ell)$ *freely generate a free subgroup of* L *with rank* $f(\ell)$. *Also, for all* $i \geq f(\ell)$ *we have* $\gamma(i,\ell) = 1$.

Proof: The final claim was verified above. At this point we sketch the proof of the first claim; a detailed proof is given below in Section F. Recall that none of $\phi(0,\ell), \ldots, \phi(f(\ell) - 1, \ell)$ is equal to 1 in L and each lies in the factor H of L. Consequently, none of $\gamma(0,\ell), \ldots, \gamma(f(\ell) - 1, \ell)$ is "obviously" equal to 1 in L. Note that $\{x^{-i}\, y^i\, x^i, i > 0\}$ are a Nielsen basis for the subgroup of $\langle x,y \rangle$ which they generate. The x's prohibit cancellation of any y's on forming the commutators $\gamma(i,\ell)$. Each $\gamma(i,\ell)$ for $i < f(\ell)$ will contain a y^i and lower powers of y, but no higher powers. Now one shows that for $i < f(\ell)$, the length of $\gamma(i,\ell)$ in L is just 4^{i+1}. Moreover $\gamma(0,\ell), \ldots, \gamma(f(\ell) - 1, \ell)$ have the cancellation properties of a Nielsen basis and hence freely generate a free group.||

The following is an immediate consequence of the properties of ϕ and

THEOREM 9: *If* $W_j \underset{G}{=} W_\ell$, *then for all* i $\gamma(i,j) = \gamma(i,\ell)$. *Also,* $W_j \underset{G}{=} W_\ell$ *if and only if* $f(j) = f(\ell)$.||

For any group K define P(K) to be the set of primes p such that K has an element of order p. Note that $P(H) = P(L) = \phi$, the empty set, sin H and L are torsion free.

Let $p_0, p_1, p_2, \ldots$ be a recursive enumeration of the odd primes in increasing order. Put

$$A_\ell = < z_0,\ldots,z_\ell;\ z_i^{p_i} = 1 \text{ for } 0 \leq i \leq \ell >.$$

Observe that the subgroup of A_ℓ generated by z_i and the subgroup generated by $z_i{}^2$ are isomorphic since p_i is an odd prime. Also $P(A_\ell) = \{p_0,\ldots,p_\ell\}$. Define

$$B_\ell = < z_0,\ldots,z_\ell, t_0,\ldots,t_\ell;\ z_i^{p_i} = 1,\ t_i^{-1} z_i t_i = z_i{}^2\ (0 \leq i \leq \ell) >.$$

Then B_ℓ is a strong Britton extension of A_ℓ with respect to the stable letters $\{t_0,\ldots,t_\ell\}$. By Corollary II-2, $P(B_\ell) = P(A_\ell) = \{p_0,\ldots,p_\ell\}$. Moreover, by Lemma II-4 the stable letters $\{t_0,\ldots,t_\ell\}$ freely generate a free subgroup of B_ℓ.

Define Π_{W_ℓ} to be the finite presentation given as follows:

Generators: The generators of L and the generators of B_ℓ.

Relations: The relations of L, the relations of B_ℓ and the relations

$$\gamma(i,\ell) = t_i \text{ for } 0 \leq i \leq \ell.$$

LEMMA 10: $\Pi_{W_\ell} \cong \Pi_{W_{f(\ell)}}$. *Moreover,* $P(\Pi_{W_\ell}) = P(\Pi_{W_{f(\ell)}}) = \{p_0,\ldots,p_{f(\ell)-1}\}$.

Proof: By Theorem 8, for $f(\ell) \leq i \leq \ell$ we have $\gamma(i,\ell) = 1$. Hence $t_i = 1$ in Π_{W_ℓ} for $f(\ell) \leq i \leq \ell$. Thus, in view of $t_i^{-1} z_i t_i = z_i^2$, it follows that $z_i = 1$ for $f(\ell) \leq i \leq \ell$. Further, by Theorem 9, the equation $\gamma(j,\ell) = \gamma(j,f(\ell))$ holds in Π_{W_ℓ} for all j. Consequently, the presentation for $\Pi_{W_{f(\ell)}}$ can be obtained from the presentation of Π_{W_ℓ} by a finite sequence of Tietze transformations. Hence, $\Pi_{W_\ell} \cong \Pi_{W_{f(\ell)}}$ and therefore $P(\Pi_{W_\ell}) = P(\Pi_{W_{f(\ell)}})$.

Now, in $\Pi_{W_{f(\ell)}}$ it follows that $\gamma(f(\ell), f(\ell)) = 1$ whence $t_{f(\ell)} = 1$ and $z_{f(\ell)} = 1$ as before. However, from Theorem 8 and the fact that the set $\{t_0,\ldots, t_{f(\ell)-1}\}$ freely generate a free subgroup of $B_{f(\ell)-1}$, it follows that $\Pi_{W_{f(\ell)}}$ is isomorphic to the free product of L and $B_{f(\ell)-1}$ with amalgamation defined by $\gamma(i,f(\ell)) = t_i$ for $0 \leq i \leq f(\ell)-1$. By properties of free products with amalgamation,

$$P(\Pi_{W_{f(\ell)}}) = P(L) \text{ union } P(B_{f(\ell)-1}) = P(B_{f(\ell)-1}) = \{p_0,\ldots,p_{f(\ell)-1}\}$$

since P(L) is empty.||

THEOREM 11: $W_m \underset{G}{=} W_n \leftrightarrow \Pi_{W_m} \cong \Pi_{W_n}$.

Proof: ($\rightarrow$) Suppose $W_m \underset{G}{=} W_n$. Then by Theorem 9, $f(m) = f(n)$. Hence, by Lemma 10, $\Pi_{W_m} \cong \Pi_{W_n}$. ($\leftarrow$) Suppose on the contrary that $W_m \underset{G}{\neq} W_n$. By Theorem 9, $f(m) \neq f(n)$. Hence, by Lemma 10, $P(\Pi_{W_m}) \neq P(\Pi_{W_n})$. Consequently, $\Pi_{W_m} \ncong \Pi_{W_n}$, a contradiction.||

This result completes the proof of the first part of Theorem 1. Recalling the construction proving Theorem 4, note that by Lemma II-9 J has a solvable word problem. So by Lemma 5, the word problem for H is Turing equivalent to the word problem for G. But H is a two-stage Britton tower over the free group F. Hence by [11], [17], or [23]:

THEOREM 12: *Let* D *be an r.e. degree. Then there exists a f.p. torsion free group* H *which is a two-stage Britton extension of a free group* F *and has word problem of degree* D.||

C. RABIN REVISITED

The proof of Theorem 1 could be completed by appealing to results of Rabin [41] and Boone [12]. Instead, we will give an alternative version of Rabin's construction which uses Britton's Lemma.

Let $\underline{M}$ be an algebraic property of finitely presented groups (i.e. $\underline{M}$ is preserved under isomorphism). $\underline{M}$ is said to be a *Markov property* if (1) there is a f.p. group $A_1 \in \underline{M}$ and (2) there is a f.p. group A_2 which is *not* isomorphic to a subgroup of any $B \in \underline{M}$. For example, any hereditary property of f.p. groups is Markov.

Rabin's Theorem: Let $\underline{M}$ be a Markov property of f.p. groups and E a fixed f.p. group. Then there is a recursive class $\theta = \{\Delta_W, W \in E\}$ of presentations of groups indexed by words of E such that $\Delta_W \in \underline{M} \leftrightarrow W = 1$ in E.

Proof: Since $\underline{M}$ is Markov, there are two f.p. groups A_1 and A_2 with properties (1) and (2). Write $E * A_2$ in presentation form as $< U_1,\ldots,U_n;\ R_1,\ldots,R_m >$ and define

$$J = < U_1,\ldots,U_n,\ x,\ y_1,\ldots,y_n,\ y_{n+1};\ R_1,\ldots,R_m,\ y_i^{-1}\ (U_i x^2) y_i = (U_i x^2)^2\ i = 1,\ldots,n,\ y_{n+1}^{-1}\ x\ y_{n+1} = x^2 >.$$

Then J is a strong Britton extension of $E * A_2 * \langle x \rangle$ with respect to the stable letters $\{y_1,\ldots,y_{n+1}\}$ since the $(U_i x^2)$ and x all have infinite order in $E * A_2 * \langle x \rangle$ and so generate subgroups isomorphic to those generated by their squares. By Lemma II-4, the y_i freely generate a free subgroup of rank $n + 1$.

Now define a f.p. group K by:

Generators: The generators of J and z.

Relations: The relations of J and $z^{-1} y_j z = y_j^2$, $j = 1,\ldots, n + 1$.

Then K is a strong Britton extension of J with respect to the stable letter z. Hence $E * A_2 * \langle x \rangle$ is embedded in K. Let $W \in E$ be a fixed word and assume $W \neq 1$ in E. Then $[W,x] = W^{-1}x^{-1}Wx \neq 1$ in J and no power of $[W,x]$ is equal in J to a word in the y_i by Britton's Lemma. Thus, in K, z and $[W,x]$ freely generate a free subgroup of rank 2 since by Britton's Lemma a non-trivial relation would imply a power of $[W,x]$ is equal in J to a word in the y_i.

Put $Q = \langle r,s,t;\ r^{-1}sr = s^2,\ s^{-1}ts = t^2 \rangle$. Then Q is a two-stage Britton extension of $\langle t \rangle$ obtained by adding s (stage 1) and then r (stage 2) as stable letters. Now r and t freely generate a free subgroup of Q of rank 2. For suppose there were a non-trivial relation between r and t. Then by Britton's Lemma at stage 2, some t^m would be equal in stage 1 to a word on s. But this is impossible, by Britton's Lemma at stage 1. Also note that Q is torsion-free.

Then assuming $W \neq 1$ we form $D_W = (K * Q;\ r = [W,x],\ t = z)$. It follows that $E * A_2 * \langle x \rangle$ is embedded in D_W. Let Δ_W be $D_W * A_1$ written out in presentation form.

Now *drop* the assumption that $W \neq 1$ in E. Then Δ_W is still a well defined recursive class of finite presentations of groups indexed by E. If $W \neq 1$ in E the above shows that A_2 is embedded in D_W and hence Δ_W so that $\Delta_W \notin \underline{M}$ by property (2).

However, if $W = 1$ in E, then $[W,x] = r = 1$ and so: $s = 1$, $t = 1$, $z = 1$, $y_i = 1$ $(1 \leq i \leq n+1)$, $x = 1$, and $U_i = 1$ $(1 \leq i \leq n)$. (Each of these equations follows from previous equations and the defining relations.) Hence $\Delta_W \cong A_1$ and so $\Delta_W \in \underline{M}$.||

Observe that if E, A_1 and A_2 are all torsion free, then so are all of the Δ_W, by properties of Britton extensions and free products with amalgamation. Moreover, if $U = V$ in E then $\Delta_U \cong \Delta_V$.

To complete the proof of Theorem 1, we apply Rabin's Theorem to the Markov property of "being trivial" – specifically, pick $A_1 = 1$, the trivial group, A_2 an infinite cyclic group. Then $\Delta_W \cong 1 \leftrightarrow W = 1$ in E.

For E we choose the group L of Section B. Then Δ_W is torsion free for $W \in L$. In the definition of Π_{W_ℓ} replace L by $\Delta_{\phi(0,\ell)}$. Then $P(\Pi_{W_\ell})$ remains unchanged and $\Pi_{W_\ell} \cong 1 \leftrightarrow \Delta_{\phi(0,\ell)} \cong 1 \leftrightarrow \phi(0,\ell) \underset{H}{=} 1 \leftrightarrow W_\ell \underset{G}{=} 1$. Since $\phi(0,\ell) = \phi(0,f(\ell))$ we still have $\Delta_{\phi(0,\ell)} \cong \Delta_{\phi(0,f(\ell))}$ and so $\Pi_{W_\ell} \cong \Pi_{W_{f(\ell)}}$. Consequently, Theorem 11 still holds and the final claim in Theorem 1 has been verified.

D. GENERALIZATIONS

Let E(i,j) be an equivalence relation on N. The direction ($\leftarrow$) of Theorem 2 is easily proved by enumerating Tietze transformations. We omit this argument. To prove ($\rightarrow$) we assume E is r.e. Let $F = \langle a_i, i \in N \rangle$ be a free group on countably many generators. Let R be the normal closure in F of the words a_0 and $a_i a_j^{-1}$ such that E(i,j). Then F/R is recursively presented.

By a theorem of Higman, Neumann, and Neumann [26], F/R can be effectively embedded in a 2-generator recursively presented group A (see

[44] for instance). By the main theorem of Higman [25], A can be effectively embedded in a f.p. group G. Let $1 = W_0, W_1, W_2, \ldots$ be the recursive list of words in G corresponding to (i.e. images of) the a_i. Now $W_i \underset{G}{=} W_j \leftrightarrow E(i,j)$. Identifying this G and list of words with those in the proof of Theorem 1 we have the desired result.

REMARKS: In [14] Boone's construction is applied to give a recursive class $\theta = \{M_W, W \in G\}$ of finite presentations of n-manifolds ($n \geq 4$) such that M_U is diffeomorphic to $M_V \leftrightarrow \Pi_U \cong \Pi_V$ where $\Omega = \{\Pi_W, W \in G\}$ is the class of groups produced by Boone's method; moreover, Π_W is the fundamental group of M_W.

Unfortunately, the construction of [14] does not yield a result corresponding to Theorem 1 for manifolds when applied to our class Ω of Theorem 1. The basic difficulty is that the diffeomorphism class of M_W constructed from Π_W is not preserved under *all* Tietze transformations of Π_W by the construction. Consequently, the second Betti number of M_W depends on the difference δ between the number of relations of Π_W and the number of generators of Π_W. For Boone's class Ω, $\delta(\Pi_W)$ is a constant so the construction works. However, for our class Ω, $\delta(\Pi_{W_\ell}) = \ell + c$ where c is a fixed constant, so the construction does not yield the desired result.

However, combining Corollary 3 of the present chapter with Corollary A, p. 41 of Boone, Haken, and Poenaru [14], the following is immediate:

THEOREM 13: *Let* D *be an r.e. many-one degree of unsolvability. Then there is a recursive class θ of finite presentations of n-manifolds* ($n \geq 4$) *such that the i-equivalence problem between members of θ has degree* D. *Here i-equivalent means one of: diffeomorphic, homeomorphic, combinatorially equivalent, or homotopy equivalent.* ||

Finally, we apply Theorem 2 to obtain the following analogue for the isomorphism problem of a theorem of Shepherdson [47] (for semi-groups):

THEOREM 14: *Let* A(x,m) *be a recursively enumerable predicate and let* D_m *be the degree of* $\{x \mid A(x,m)\}$. *Further let* D *be any r.e. degree such that degree* $A(x,m) \leq_T D$ *(so* $D_m \leq_T D$*). Then there is a recursive class* $\Omega = \{\Pi_i,\ i \in N\}$ *such that:*

(1) The isomorphism problem for Ω *has degree* D;

(2) For each n, *the problem of deciding for arbitrary* $\Pi \in \Omega$ *whether or not* $\Pi \cong \Pi_n$ *has degree one of* $\underline{O}$, D_1, $D_2, \ldots$. *(This is the special isomorphism problem for* Π_n*);*

(3) There is a recursive function h *such that for each* k, $\Pi_{h(k)}$ *has special isomorphism problem of degree* D_k.

In particular, if all $D_i = \underline{O}$ *and* $D > \underline{O}$, *then the special isomorphism problem for each* $\Pi \in \Omega$ *is solvable, but not uniformly since* Ω *has isomorphism problem of degree* $D > \underline{O}$.

Proof: In view of Theorem 2, we need only produce an equivalence relation with the correct properties. Let g be a recursive function such that range $g \in D$. Define an r.e. predicate B(z,m) as follows:

$$B(z,m) \quad \leftrightarrow \quad \begin{cases} (1)\ z = 1 \text{ or} \\ (2)\ z = 2 \text{ or} \\ (3)\ z = x + 3 \\ \qquad \text{and } A(x,m) \text{ holds.} \end{cases}$$

Then B(O,m) never holds, while B(1,m) and B(2,m) always holds. Moreover, A and B have the same degree and $\{z \mid B(z,m)\} \in D_m$.

Let $p_o, p_1, \ldots$ be the (recursive) sequence of all odd primes. Define $E(k,\ell)$ as follows:

$$E(k,\ell) \leftrightarrow \begin{cases} (1)\ k = \ell \text{ or} \\ (2)\ \text{for some } n,\ k = p_{g(n)}^{a} \text{ and } \ell = p_{g(n)}^{b} \text{ and} \\ \quad B(a,n) \text{ and } B(b,n) \text{ both hold.} \end{cases}$$

Then $E(k,\ell)$ is clearly an r.e. equivalence relation. From the properties of B, it follows that $E(p_i, p_i^2)$ holds $\leftrightarrow$ $i \in$ range g. Since degree $B \leq_T D$, this implies that $E(k,\ell)$ has degree D.

Now $E(p_{g(n)}^{a}, p_{g(n)}^{b})$ holds $\leftrightarrow$ both B(a,n) and B(b,n) hold. From the definition of E, it follows that the problem of whether or not $E(p_{g(n)}^{a}, \ell)$ holds for arbitrary ℓ has degree D_n if B(a,n) holds or has degree $\underline{O}$ if B(a,n) does not hold. Also note that if k does not have form $p_{g(n)}^{a}$ where B(a,n) holds for some n, then $E(k,\ell)$ holds $\leftrightarrow k = \ell$. Thus for such k whether or not $E(k,\ell)$ holds has degree $\underline{O}$.

Let $\Omega = \{\Pi_i,\ i \in N\}$ be the class of groups constructed from E via Theorem 2. Define $h(n) = p_{g(n)}$. All of the assertions of the theorem are now easily verified by using the above discussion concerning E.||

E. ISOMORPHISM PROBLEMS OF ARBITRARY RECURSIVELY ENUMERABLE MANY-ONE DEGREE

Notice that Corollary 3 is trivial for the many-one degree of the empty set – just pick the empty class of finite presentations. For other many-one degrees Corollary 3 follows immediately from Theorem 2 and the following Lemmas:

LEMMA 15: *Every r.e. many-one degree (except that of the empty set) contains an r.e. equivalence relation on the natural numbers* N.

Proof: Since every set in an r.e. many-one degree D is r.e. (see [43]) we need only show D contains an equivalence relation. Without loss of generality we may pick $A \in D$ where A is infinite and $N \setminus A$ is non-empty. Define E(i,j) to be the equivalence relation consisting of all ordered pairs (i,i) where $i \in N$, (2j, 2j+1) and (2j+1, 2j) where $j \in A$. Notice that each equivalence class consists of one or four ordered pairs. Clearly, A is many-one reducible to E (even one-one reducible). It remains to show that E is many-one reducible to A. Let $k \in N \setminus A$ and $\ell \in A$ be fixed. Define a recursive function f on ordered pairs of natural numbers to N by:

$$f((m,n)) = \begin{cases} \ell \text{ if } (m,n) \text{ has form } (i,i) \\ j \text{ if } (m,n) \text{ has form } (2j,\, 2j+1) \text{ or } (2j+1,\, 2j) \\ k \text{ otherwise.} \end{cases}$$

Then $(m,n) \in E \leftrightarrow f((m,n)) \in A$. So E is many-one reducible to A.||

LEMMA 16: *The one-one degree of a simple set* A *does not contain an equivalence relation.*

Proof: Recall that A is r.e. and being r.e. and simple are preserved under one-one equivalence. Suppose on the contrary there is an equivalence relation (necessarily r.e. and simple) $E \equiv_1 A$. Since E is not recursive, it defines at least two equivalence classes, say X and Y. Pick two fixed representatives $x_0 \in X$ and $y_0 \in Y$. Assume Y is infinite. Then $\{(x_0,z) : E(z,y_0)\}$ is an infinite r.e. subset of $\overline{E}$ – a contradiction. Hence Y (indeed all classes) must be finite. Let $Y = \{y_0, y_1, \ldots, y_n\}$. Then the set $\{(y_0,z) : z \neq y_0, z \neq y_1, \ldots, z \neq y_n\}$ is an r.e. subset of $\overline{E}$ – again a contradiction.||

The author is indebted to Professor Carl Jockusch for Lemma 15 and to Professor K. I. Appel for Lemma 16.

F. SUPPLEMENT ON CERTAIN COMMUTATORS

This section is entirely devoted to giving a detailed proof of the remaining part of Theorem 8: $\gamma(0,\ell),\ldots,\gamma(f(\ell)-1,\ell)$ freely generate a free subgroup of L of rank $f(\ell)$. Without loss of generality we assume $f(\ell) > 0$. (Otherwise the assertion is trivial.)

By definition of f and ϕ, none of $\phi(0,\ell),\ldots,\phi(f(\ell)-1,\ell)$ is equal to 1 in L. Since $f(\ell) > 0$, $\gamma(0,\ell) = [x^{-1}\,y\,x,\ \phi(0,\ell)]$ is not equal to 1 in L and has length 4. Consequently, without loss of generality we assume hereafter that $f(\ell) > 1$.

LEMMA 17: *Suppose* $i < f(\ell)$. *Further assume: length* $\gamma(i-1,\ell) = d \geq 4$ *and for* $\gamma(i-1,\ell)$ *in normal form:*

(a) $\gamma(i-1,\ell)$ *begins with* $x^{-i}\,y^{-i}$

(b) $\gamma(i-1,\ell)$ *ends in a word* $U \in H$ *with* $u \neq 1$

(c) y^i *occurs in* $\gamma(i-1,\ell)$

(d) *the powers of* x *and* y *in* $\gamma(i-1,\ell)$ *are* $\leq i$.

Then: $\gamma(i,\ell) \neq 1$, *length* $\gamma(i,\ell) = 4d$ *and for* $\gamma(i,\ell)$ *in normal form:*

(a^+) $\gamma(i,\ell)$ *begins with* $x^{-(i+1)}\,y^{-(i+1)}$.

(b^+) $\gamma(i,\ell)$ *ends in a word* $y^{i+1}\,x^{i+1}\,V$, $1 \neq V \in H$.

(c^+) y^{i+1} *occurs in* $\gamma(i,\ell)$.

(d^+) *the powers of* x *and* y *in* $\gamma(i,\ell)$ *are* $\leq i+1$.

Proof: Let $\beta = [\gamma(i-1,\ell),\ x^{-(i+1)}\,y^{i+1}\,x^{i+1}]$. Then

$$\beta = \gamma(i-1,\ell)^{-1}\,x^{-(i+1)}\,y^{-(i+1)}\,x^{i+1}\,\gamma(i-1,\ell)\,x^{-(i+1)}\,y^{i+1}\,x^{i+1},$$

where $\gamma(i-1,\ell)$ is assumed to be in normal form. Observe that between $\gamma(i-1,\ell)^{-1}$ and $x^{-(i+1)}\,y^{-(i+1)}x^{i+1}$ and between $x^{-(i+1)}y^{-(i+1)}\,x^{i+1}$ and $\gamma(i-1,\ell)$ not all of $x^{\pm(i+1)}$ can cancel by (a). Between $\gamma(i-1,\ell)$ and $x^{-(i+1)}\,y^{i+1}\,x^{i+1}$ no cancellation occurs by (b). Thus length $\beta = 2d$. Now β ends in $y^{i+1}\,x^{i+1}$ and begins in $U^{-1} \in H$, $U \neq 1$ by (b). The powers of x and y in β are $\leq i+1$. Now $\gamma(i,\ell) = [\beta, \phi(i,\ell)] = \beta^{-1}\phi(i,\ell)^{-1}\,\beta\,\phi(i,\ell)$. Assuming β in normal form, by the properties of β it follows that length $\gamma(i,\ell) = 2(2d) = 4d$ – note that around $\phi(i,\ell)^{-1}$ we have H-part $U\phi(i,\ell)^{-1}U^{-1} \neq 1$. Properties (a^+), (b^+), (c^+) and (d^+) follow from the properties of β and $\phi(i,\ell) \neq 1$.||

COROLLARY 18: *Suppose* $i < f(\ell)$. *Then* $\gamma(i,\ell) \neq 1$, *length* $\gamma(i,\ell) = 4^{i+1}$, *and properties* (a^+) *thru* (d^+) *of Lemma 17 hold for* $\gamma(i,\ell)$.

Proof: The claim is immediate for $\gamma(0,\ell)$. The result follows for $i < f(\ell)$ by induction and Lemma 17.||

COROLLARY 19: *Suppose* $m, n < f(\ell)$ *and* $m \neq n$. *Then*

(i) *length* $\gamma(m,\ell)\gamma(n,\ell) = 4^{m+1} + 4^{n+1}$.

(ii) *length* $\gamma(m,\ell)^{-1}\gamma(n,\ell) = 4^{m+1} + 4^{n+1} - 1$.

(iii) *length* $\gamma(m,\ell)\gamma(n,\ell)^{-1} \geq 4^{m+1} + 4^{n+1} - 3$.

Proof: By Corollary 18, properties (a^+) thru (d^+) of Lemma 17 apply to $\gamma(m,\ell)$ and $\gamma(n,\ell)$ which have lengths 4^{m+1} and 4^{n+1} respectively. (i) is immediate by (a^+) and (b^+). (ii) follows by (a^+) since $m \neq n$ so $x^{m+1}\,x^{-(n+1)} \neq 1$. (iii) follows by (b^+) since $m \neq n$ so in $\gamma(m,\ell)\gamma(n,\ell)^{-1}$ the subword (at juxtaposition)

$$y^{m+1}\,x^{m+1}\,V_m\,V_n^{-1}\,x^{-(n+1)}\,y^{-(n+1)}$$

still leaves x's uncancelled.||

COROLLARY 20: $\gamma(0,\ell),\ldots,\gamma(f(\ell)-1,\ell)$ *freely generate a free subgroup of rank* $f(\ell)$.

Proof: By Corollary 19, no $\gamma(i,\ell)$ $(i < f(\ell))$ is absorbed in any others when words are formed. Thus the listed words have the properties of a Nielsen basis.||

Corollary 20 completes the proof of Theorem 8.

LIST OF REFERENCES

[1] S. I. Adjan, *On algorithmic problems in effectively complete classes of groups* (Russian), Doklady Akad. Nauk SSSR, 123 (1958), 13-16.

[2] G. Baumslag, *A remark on generalized free products*, Proc. of Amer. Math. Soc., 13 (1962), 53-54.

[3] G. Baumslag, *Automorphism groups of residually finite groups*, Journ. London Math. Soc., 38 (1963), 117-118.

[4] G. Baumslag, *Finitely presented groups*, Proc. Internat. Conf. Theory of Groups, Australian National University, Canberra, August 1965, pp 37-50, Gordon and Breach, New York, 1967.

[5] G. Baumslag, W. W. Boone, and B. H. Neumann, *Some unsolvable problems about elements and subgroups of groups*, Math. Scand., 7 (1959), 191-201.

[6] N. Blackburn, *Conjugacy in nilpotent groups*, Proc. of Amer. Math. Soc., 16 (1965), 143-148.

[7] L. A. Bokut', *On a property of the groups of Boone* (Russian), Algebra i logika, Seminar, vol. 5. no. 5 (1966), 5-23, and vol. 6 no. 1 (1967), 15-24.

[8] W. W. Boone, *Certain simple unsolvable problems of group theory*, Indig. Math. 16 (1954), 231-237, 492-497; 17 (1955), 252-256, 571-577; 19 (1957), 22-27, 227-232.

[9] W. W. Boone, *The word problem*, Ann. of Math., 70 (1959), 207-265.

[10] W. W. Boone, *Partial results regarding word problems and recursively enumerable degrees of unsolvability,* Bull. of Amer. Math. Soc., 68 (1962), 616-623.

[11] W. W. Boone, *Word problems and recursively enumerable degrees of unsolvability. A sequel on finitely presented groups,* Ann. of Math., 84 (1966), 49-84.

[12] W. W. Boone, *Decision problems about algebraic and logical systems as a whole and recursively enumerable degrees of unsolvability,* Contributions to Mathematical Logic, K. Schütte, editor, North-Holland, Amsterdam, 1968.

[13] W. W. Boone, *The theory of decision processes in group theory: a survey,* Amer. Math. Soc. invited address, to be published in Bull. of Amer. Math. Soc.

[14] W. W. Boone, W. Haken, and V. Poenaru, *On recursively unsolvable problems in topology and their classification,* contributions to Mathematical Logic, K. Schütte, editor, North-Holland, Amsterdam, 1968.

[15] W. W. Boone and H. Rogers, Jr., *On a problem of JHC Whitehead and a problem of Alonzo Church,* Math. Scand. 19 (1966), 185-192.

[16] J. L. Britton, *The word problem,* Ann. of Math., 77 (1963), 16-32.

[17] C. R. J. Clapham, *Finitely presented groups with word problems of arbitrary degrees of insolvability,* Proc. London Math. Soc., Series 3, 14 (1964), 633-676.

[18] D. J. Collins, *Recursively enumerable degrees and the conjugacy problem,* Acta Mathematica, Vol. 122 (1969), 115-160.

[19] D. J. Collins, *On embedding groups and the conjugacy problem,* to be published.

[20] M. Dehn, *Uber unendliche diskontinuierliche Gruppen*, Math. Annalen, 71 (1911), 116-144.

[21] V. H. Dyson, *The word problem and residually finite groups*, Notices of Amer. Math. Soc., 11 (1964), 743.

[22] A. A. Fridman, *On the relation between the word problem and the conjugacy problem in finitely defined groups* (Russian), Trudy Moskov, Mat. Obsc., 9 (1960), 329-356.

[23] A. A. Fridman, *Degrees of unsolvability of the word problem for finitely defined groups* (Russian), Izdalel'stvo "Nauka", Moscow 1967, 193 pp.

[24] M. Hall, *Subgroups of finite index in free groups*, Canadian Journ. Math., 1 (1949), 187-190.

[25] G. Higman, *Subgroups of finitely presented groups*, Proc. of Royal Soc., A, vol. 262 (1961), 455-475.

[26] G. Higman, B. H. Neumann, and H. Neumann, *Embedding theorems for groups*, Journ. London Math. Soc., 24 (1949), 247-254.

[27] A. G. Kurosh, *The theory of groups*, Vol. II, Chelsea, New York, 1956.

[28] R. C. Lyndon, *Groups with parametric exponents*, Trans. of Amer. Math. Soc., 96 (1960), 518-533.

[29] R. C. Lyndon, *On Dehn's algorithm*, Math. Annalen 166 (1966), 208-228.

[30] W. Magnus, A. Karrass, and D. Solitar, *Combinatorial group theory*, Wiley, New York, 1966.

[31] A. I. Mal'cev, *On homomorphisms on finite groups*, (Russian), Ucen. Zap. Ivanousk. Gos. Ped. Inst., 18 (1958), 49-60.

[32] C. F. Miller, III, *On Britton's theorem A,* Proc. of Amer. Math. Soc., 19 (1968), 1151-1154.

[33] C. F. Miller, III and P. E. Schupp, *The geometry of Britton extensions and free products with amalgamation,* in preparation.

[34] C. F. Miller, III and P. E. Schupp, *Embedding into hopfian groups,* Journ. of Algebra, 17 (1971) 171-176.

[35] A. W. Mostowski, *On the decidability of some problems in special classes of groups,* Fund. Math., LIX (1966), 123-135.

[36] V. L. Murskij, to be published.

[37] B. H. Neumann, *An essay on free products of groups with amalgamations,* Phil. Trans. Royal Soc. of London, No. 919, 246 (1954), 503-554.

[38] H. Neumann, *Varieties of groups,* Ergebnisse der Mathematik und ihrer Grenzgebiete, Band 37, Springer-Verlag, Berlin-Heidelberg, 1967.

[39] P. S. Novikov, *Unsolvability of the conjugacy problem in the theory of groups* (Russian), Izv. Akad. Nauk SSSR, Ser. Mat., 18 (1954), 485-524.

[40] P. S. Novikov, *On the algorithmic unsolvability of the word problem in groups* (Russian), Trudy Mat. Inst. Steklov no. 44. Izdat Akad. Nauk SSSR, Moscow, 1955.

[41] M. O. Rabin, *Recursive unsolvability of group theoretic problems,* Ann. of Math., 67 (1958), 172-194.

[42] K. Reidemeister, *Einführung in die Kombinatorische Topologie,* Braunschweig, 1932. (also Chelsea, New York, 1950)

[43] H. Rogers, Jr., *Theory of recursive functions and effective computability,* McGraw-Hill, New York, 1967.

[44] J. J. Rotman, *The theory of groups,* Allyn and Bacon, Boston, 1965.

[45] P. E. Schupp, *On Dehn's algorithm and the conjugacy problem,* Math. Annalen, 178 (1968), 119-130.

[46] H. Seifert and W. Threlfall, *Lehrbuch der topologie,* Teubner, Leipzig, 1934. (Also Chelsea, New York, 1947.)

[47] J. C. Shepherdson, *Machine configuration and word problems of given degree of unsolvability,* Zeit. f. Math. Logik und Grund. d. Math., 11 (1965), 149-175.

[48] J. R. Shoenfield, *Mathematical logic,* Addison-Wesley, Reading, Mass. 1967.

[49] E. H. Spanier, *Algebraic topology,* McGraw-Hill, New York, 1966.

[50] V. N. Remeslennikov, *Conjugacy in polycyclic groups* (Russian), Algebra: Logika, Seminar, Vol. 8., No. 6 (1969), 712-725.

[51] P. Hall, *Finiteness conditions for soluble groups,* Proc. London Math. Soc., (3) 4 (1954), 419-436.

[52] K. A. Mihailova, *The occurrence problem for direct products of groups,* Dokl. Akad. Nauk SSSR, 119 (1958), 1103-1105 (Russian).

[53] C. F. Miller, III, *On group-theoretic decision problems and their classification,* Thesis, University of Illinois, 1969.

[54] C. F. Miller, III, *Decision problems in algebraic classes of groups (a survey),* Proceedings of the Irvine Conference on Decision Problems in Group Theory, Studies in Logic, North-Holland, Amsterdam, to be published.

[55] I. N. Sanov, *A property of a certain representation of a free group* (Russian), Doklady Akad. Nauk SSSR, vol. 57 (1947), 657-659.

[56] P. E. Schupp, *A survey of small cancellation theory,* Proceedings of the Irvine Conference on Decision Problems in Group Theory, Studies in Logic, North-Holland, Amsterdam, to be published.

[57] K. H. Toh, *Problems concerning residual finiteness in nilpotent groups,* to be published.

[58] D. J. Collins, *Representation of Turing reducibility by word and conjugacy problems in finitely presented groups,* to appear in Acta Mathematica.

INDEX OF SYMBOLS

Note: $\neq$ means "not =" etc.

INDEX